P9-CSC-895

ANCIENT CULTURE AND SOCIETY

GREEK SCIENCE AFTER ARISTOTLE

ANCIENT CULTURE AND SOCIETY

General Editor
M. I. FINLEY

Professor of Ancient History
at the University of Cambridge

A. W. H. ADKINS Moral Values and Political Behaviour in Ancient Greece
H. C. BALDRY The Greek Tragic Theatre
P. A. BRUNT Social Conflicts in the Roman Republic
M. I. FINLEY Early Greece: the Bronze and Archaic Ages
A. H. M. JONES Augustus
G. E. R. LLOYD Early Greek Science: Thales to Aristotl(
G. E. R. LLOYD Greek Science After Aristotle
C. MOSSÉ The Ancient World at Work
R. M. OGILVIE The Romans and their Gods
B. H. WARMINGTON Nero: Reality and Legend

Other titles in preparation

GREEK SCIENCE AFTER ARISTOTLE

G. E. R. LLOYD

Senior Tutor of King's College, Cambridge

W · W · NORTON & COMPANY · INC · NEW YORK

Printed in the United States of America.

Library of Congress Cataloging in Publication Data
Lloyd, Geoffrey Ernest Richard.
Greek science after Aristotle.
(Ancient culture and society)
Bibliography: p.
1. Science–History–Greece. 2. Technology–History–Greece. I. Title.
Q127.G7L58 509'.38 72-11959
ISBN 0-393-04371-1
ISBN 0-393-00689-1 (pbk)

1 2 3 4 5 6 7 8 9 0

CONTENTS

FIGURES

FIGURES

Note on Pronunciation

In Greek the letter *e* is always a sign of a new syllable, unlike the *e* in *bone*. It is either short, as in *met*, or long, in which case it is pronounced as the *ê* in the French *tête*. In the Greek words mentioned in this book the vowel is short unless otherwise marked.

ACKNOWLEDGEMENTS

THE author and publishers are grateful to the Clarendon Press, Oxford for permission to reproduce five illustrations from *A History of Technology* volumes 2 and 3 (Figs. 15, 18, 20, 24 and 26); two illustrations from *Grain-Mills and Flour in Classical Antiquity* by L. A. Moritz (Figs. 22 and 23); one illustration from *Greek and Roman Artillery, Historical Development* by E. W. Marsden (Fig. 14). Also to Harvard University Press for permission to reproduce one illustration from *A Source Book in Greek Science* by M. R. Cohen and I. E. Drabkin (Fig. 17) and Munksgaard for one illustration from *The Mechanical Technology of Greek and Roman Antiquity* by A. G. Drachmann (Fig. 16). Figs. 11, 19 and 21 are reproduced from *Heronis Alexandrini, Opera* I and III edited by W. Schmidt and H. Schöne, (Leipzig 1899, 1903). The remaining diagrams and the map have been drawn by F. Rodney Fraser.

CHRONOLOGICAL TABLE

Only the most important scientists down to Galen are listed (the dates of later scientists are noted when their work is discussed in Chapter 10). The precise dates of birth and death of ancient scientists are not often known: where they are, they are given in brackets. Otherwise the date opposite a scientist's name is intended merely as a rough guide to the period when he may be presumed to have done his chief work.

Scientists			*Contemporary events*
		B.C.	
Aristotle of Stagira (384–322)			
		323	Death of Alexander
Theophrastus of Eresus (371–286)			
		304	Ptolemy I Soter king of Egypt
Praxagoras of Cos	300		
Euclid	300		
Epicurus of Athens (341–270)			
Zeno of Citium (335–263)			
Strato of Lampsacus	290		
		285	Ptolemy II Philadelphus joint ruler
Cleanthes of Assus (331–232)			
Aristarchus of Samos	275		
Ctesibius of Alexandria	270		
Herophilus of Chalcedon	270		
		269	Hiero II king of Syracuse
Erasistratus of Ceos	260		
		246	Ptolemy III Euergetes succeeds

CHRONOLOGICAL TABLE

Scientists			*Contemporary events*
		B.C.	
Archimedes of Syracuse (287–212)			
Chrysippus of Soli (280–207)			
Eratosthenes of Cyrene	225		
		221	Ptolemy IV Philopator succeeds
		216	Battle of Cannae
		212	Romans take Syracuse
Apollonius of Perga	210		
Philo of Byzantium	200		
		168	Battle of Pydna
Seleucus of Seleucia	150		
		146	Rome destroys Carthage, and Corinth
		145	Ptolemy VIII Euergetes II Physcon
Hipparchus of Nicaea	135		
		133	Kingdom of Pergamum bequeathed to Rome
		86	Sulla sacks Athens
Posidonius of Apamea	80		
Lucretius	60		
		48	Bellum Alexandrinum
		31	Battle of Actium
Vitruvius	25		
		A.D.	
Strabo of Amasia	10		
		14–37	Tiberius emperor
Celsus	40		
Hero of Alexandria	60		
		69–79	Vespasian emperor
Menelaus of Alexandria	95		
		98–117	Trajan emperor
Rufus of Ephesus	100		
		117–138	Hadrian emperor

CHRONOLOGICAL TABLE

Scientists		*Contemporary events*	
		A.D.	
Soranus of Ephesus	120		
		138–161	Antoninus Pius emperor
Ptolemy of Alexandria	150		
		161–180	Marcus Aurelius emperor
Galen of Pergamum	180		

PREFACE

THIS book is a continuation of my *Early Greek Science: Thales to Aristotle.* Several of the preliminary points that I emphasized in that study must be repeated at the outset. First, science is a modern category, not an ancient one. There is no one term, in Greek or Latin, that is exactly equivalent to our 'science'. The terms in which the ancients themselves describe what we should call their scientific work include, for example, *peri physeōs historiā* (inquiry concerning nature), *philosophiā* (love of wisdom, philosophy), *theōriā* (speculation) and *epistēmē* (knowledge), and different ancient authors have quite different conceptions of the nature, aims and methods of the inquiry they were undertaking. Thus a good deal of what we know as early Greek science is embedded in philosophy, and this remains true, though to a lesser degree, of the period after Aristotle. 'Physics' is treated as one of the three branches of philosophy by the Hellenistic[1] philosophers, the other two being 'logic' and 'ethics', and the commonest justification offered for the inquiry concerning nature is a philosophical one, that it contributes to wisdom. On the other hand some other writers, including mathematicians and doctors especially, either ignore philosophy or explicitly dissociate themselves from the philosophers.

As in my earlier study, we shall be concerned both with the theories, problems and methods of the particular branches of science that engaged the attention of the ancient authors, and with their ideas on the nature of the scientific inquiry itself. The source material available to us, both literary and—in the case of applied science and technology—archaeological, is much richer than for the period down to Aristotle. There are, to be sure, large gaps in our evidence, especially concerning some of the Hellenistic biologists and astronomers. Nevertheless we have a considerable body of texts on which to base our

[1] The term Hellenistic is used conventionally to refer to the period that runs roughly from the death of Alexander the Great (323 B.C.) to the end of the Ptolemaic dynasty and the Roman annexation of Egypt (30 B.C.).

study, including a substantial proportion of the work of such men as Euclid, Archimedes, Ptolemy and Galen, as well as many lesser figures.

Our treatment of this material must necessarily be highly selective in a work of this scope. Only a minute proportion of the specific problems and theories of the ancient writers can be mentioned. We shall devote more attention, proportionately, to the evidence that bears on their views on the nature of the inquiries on which they were engaged. The differing and competing views of the scientific investigation itself, and the interrelations of science and philosophy, science and religion, and science and technology, provide the central themes of our discussion and it is with these general and fundamental topics chiefly in mind that I have selected the material we shall consider. In the first six chapters we shall be concentrating on the Hellenistic period, more specifically on the work done in the 200 years following the death of Aristotle (322 B.C.). The discussion of the relations between science and technology will, however, involve considering later evidence as well. Two scientists of the second century A.D., Ptolemy and Galen, are important enough to merit attention in separate chapters. A final chapter is devoted to a brief consideration of some later writers and to discussing some of the difficulties raised by the question of the decline of ancient science.

* * *

My thanks are due to friends and colleagues from various disciplines who have been most generous with their help and advice. Dr J. G. Landels, Dr V. A. Nutton, Dr R. Sibson and Mr K. D. White have read and commented on particular chapters. I have learnt much on medical matters from my father, Dr W. E. Lloyd. Both Professor M. I. Finley and Mr A. C. Reynell have read the whole book in draft and been responsible for many improvements in style and in content. To Professor Finley, who has once again advised me at each stage in the writing of the book, I owe a special debt of gratitude. It is a pleasure to express my deep appreciation for their help.

G.E.R.L.

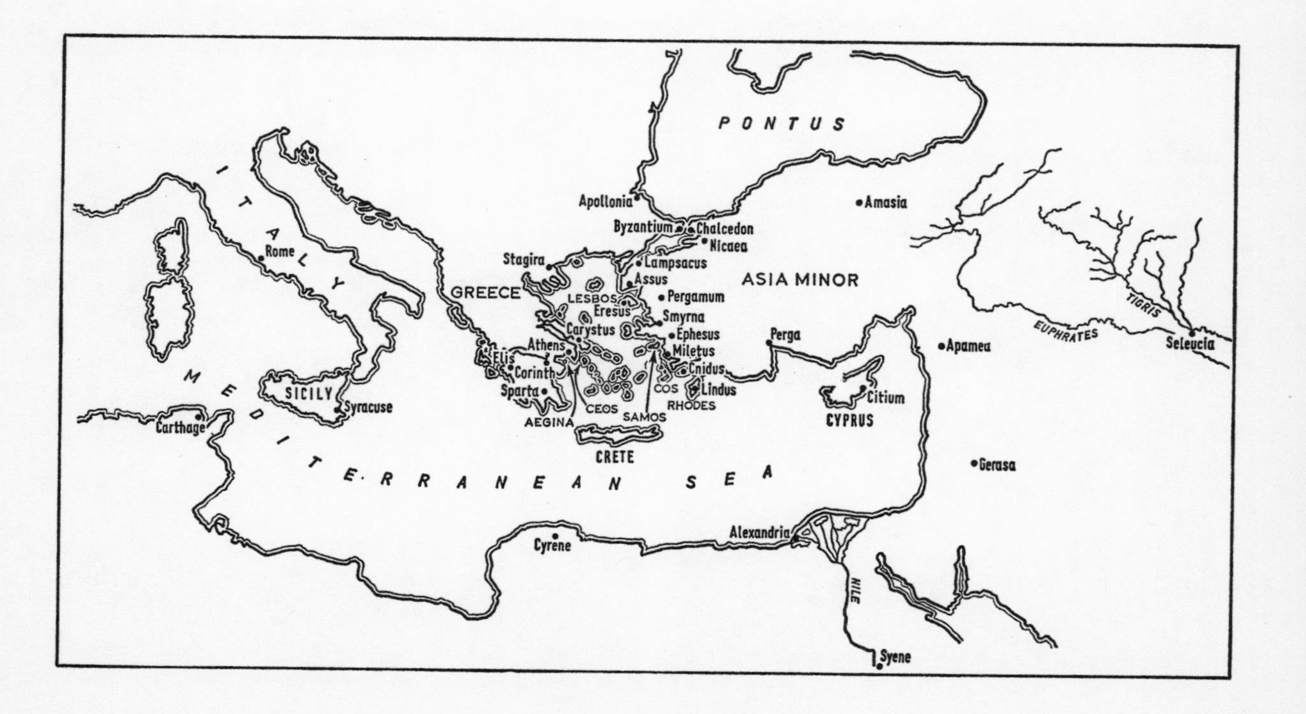
PONTUS
ITALY
Rome
SICILY
Syracuse
Carthage
MEDITERRANEAN SEA
GREECE
Stagira
Apollonia
Byzantium
Chalcedon
Nicaea
Lampsacus
Assus
ASIA MINOR
Amasia
LESBOS
Eresus
Pergamum
Smyrna
Carystus
Ephesus
Athens
Miletus
Elis
Corinth
Cnidus
Sparta
COS
Lindus
CEOS
RHODES
AEGINA
SAMOS
CRETE
Perga
Citium
CYPRUS
Apamea
Gerasa
TIGRIS
EUPHRATES
Seleucia
Cyrene
Alexandria
NILE
Syene

ANCIENT CULTURE AND SOCIETY

GREEK SCIENCE AFTER ARISTOTLE

I

Hellenistic Science: the Social Background

THE conquests of Alexander the Great brought about fundamental changes in the Greek world. His empire united a vast area under a single ruler, and, even though this unity did not survive his death in 323 B.C., his conquests had transformed the scale on which political and military operations were conducted by the Greeks. Whereas the history of the fifth and fourth centuries is largely one of the struggle for hegemony between the city-states of the Greek mainland, the arena of third-century politics extends far beyond the frontiers of the Greek-speaking peoples. The city-states retained a certain independence, but power lay not in Athens or Sparta, but in the capitals of the kingdoms carved out of Alexander's empire by his successors, the Antigonids, the Seleucids and the Ptolemies. The territories that these kingdoms controlled far surpassed those of the fifth- or fourth-century city-states, Athens included, and the revenues of those territories produced a surplus of wealth that could be devoted to whatever purposes—peaceful or military—the kings chose. The most important of these kingdoms, for our purposes, was that of the Ptolemies in Egypt. They are known to have been particularly ruthless in their exploitation of the lands they possessed, and they also happen to have been the keenest patrons of the arts, literature, scholarship and science.

The ideological changes that accompanied the changes in the power-political situation after Alexander's conquests are more difficult to describe. First, some at least of the barriers between Greeks and non-Greeks were eroded. The extent to which this happened in the third century has, to be sure, frequently been exaggerated in the past. Alexander himself has often been represented as the first great internationalist statesman, the first person to have set out to unify not merely the states, but also the peoples, of the world. Yet although he evidently wished to bring the Persian nobility into partnership with the Macedonians as the rulers of his empire and is reported to have tried to strengthen this alliance by intermarriage there is now little support for the hypothesis that he sought to extend such a policy to include other nations as well. As for his motives, the move to treat Persians on a par with the Macedonians can be explained on the grounds of political expediency—the need to broaden the basis of the ruling élite—without any appeal to a desire to unify mankind. Moreover the idea that all men share a common humanity, so far from being an entirely new conception in the third century, is one whose roots can be traced back into the fifth.

Nevertheless one of the results of Alexander's conquests was that closer intellectual and cultural contacts between Greeks and Barbarians became possible, and a certain widening of the mental horizons can be remarked in a number of third-century writers. At the same time the Hellenistic age was no more peaceful than earlier periods, and an anxiety about the uncertainties of human life is a prominent feature of the main philosophies of the time. Both Epicureanism and Stoicism have been described as refugee philosophies, and with good reason. Whereas positive conceptions of happiness are central to the moral philosophies of Plato and Aristotle, both Epicureans and Stoics attached great importance to a negative aim, the attainment of freedom from anxiety and fear. Both Hellenistic philosophies taught that happiness is independent of external factors. The wise man is unaffected by whatever apparent evil happens to him; he is happy even on the rack. Tranquillity of mind is the objective, and every activity, including—as we shall see—the investigation of nature, is subordinated to it.

The main political, economic and cultural changes that mark out what we call the Hellenistic from the classical period affected the development of science in a variety of ways. First and most obviously we are dealing no longer with purely Greek science, but with the science of the Hellenistic world. Zeno, the founder of Stoicism, came from Citium in Cyprus, a city with a strong Phoenician element in the population, and although the texts that imply that he was himself a Phoenician by birth are open to doubt, the philosophy he founded certainly numbered many non-Greeks among its adherents. Seleucus, the only astronomer we hear of who followed Aristarchus of

Samos in adopting the heliocentric theory, was a Chaldaean or Babylonian, a native of Seleucia on the Tigris. From the third century onwards the cross-fertilization between Babylonian and Greek astronomy is increasingly important: in the next century Hipparchus, for example, evidently had access to Babylonian eclipse records and used them extensively.

A second and more important distinguishing feature of the social background to Hellenistic science is the increase in kingly patronage. Science—or rather certain scientists—received substantial support, both in terms of money and, as we shall see, in other ways, from some of the Hellenistic kings, especially the Ptolemies and the Attalid dynasty of Pergamum. Thanks very largely to the Ptolemies, Alexandria, which had been founded in the year 331 B.C., rapidly became the chief centre of scientific research in the third century.

Two institutions, the Library and the Museum, stimulated this development. The Library was very probably established by Ptolemy I (Soter), and some scholars put the founding of the Museum too in his reign, although this is more commonly dated to the beginning of the reign of Ptolemy II (Philadelphus) about 280 B.C. Like Plato's Academy and Aristotle's Lyceum, the Museum was a community of scholars working, and to some extent living, together: there were common meals for the members of the Museum, as there were in the Academy and Lyceum. But the Museum differed from them first in that it was not—at least not primarily—a teaching institution, but one devoted to research, and secondly in that, whereas the Academy and the Lyceum were self-supporting, the Museum—and the Alexandrian Library too—were maintained entirely by funds provided by the Ptolemies. They not only built the Library and Museum and the complex of buildings associated with them in the royal quarter of Alexandria, but also paid regular stipends to such officials as the Librarian himself and grants to other scholars.

The principal subject that benefited from the cultural policies of the Ptolemies was not science but literature, both the writing of original works and, to an even greater extent, the scholarly study of literature. Philology, if not literary criticism itself, may be said to have begun in Alexandria. But natural science and mathematics were also beneficiaries. In some cases we hear of distinguished scientists who held official posts. Thus Eratosthenes of Cyrene, whose interests included geography, mathematics and astronomy, as well as history and literary criticism, was head of the Library at the end of the third century. On other occasions the support the Ptolemies gave was not merely financial. As we shall see later (pp 75 ff), it was only in Alexandria that the dissection of human bodies was undertaken on any scale in antiquity, and the anatomists concerned, Herophilus of Chalcedon and Erasistratus of Ceos, clearly depended on the Ptolemies' support. In the preface to book I of his work *On Medicine*, Celsus tells us that Herophilus and Erasistratus obtained the bodies of criminals from the state prisons for their investigations.

But if the fact that the Ptolemies aided certain types of scientific research is well established, the question of their motives needs clarifying. A text in Philo of Byzantium throws some light both on the variety of work they supported and on their reasons for doing so. In his book *On Artillery Construction* (ch 3), composed about 200 B.C., Philo speaks of the research carried out on the mechanical problems connected with the construction of artillery. He contrasts the crude trial-and-error methods of earlier workers with the systematic experimental investigations of the engineers at Alexandria, and, commenting on the successes they had achieved in this field, he remarks that they 'received considerable support' from 'kings who were eager for fame and well-disposed to the arts and crafts'.

One important point that this report makes clear is that the Ptolemies occasionally supported what we should call applied, as well as pure, science. In the particular field mentioned by Philo, where theoretical knowledge could be applied to the problem of perfecting weapons of war, there was a powerful practical motive for prompting scientific research. Even so research that had such direct applications was the exception, and that practical considerations were not the only ones at work, even in this instance, is suggested by Philo's reference to the Ptolemies' love of *fame*.

In many cases the chief reasons for encouraging scientists to come and work in Alexandria was simply the prestige that attached to their research. Many educated Greeks had at least some idea of the intellectual skill involved in the solution of a complex mathematical or physical problem. Although such mathematicians as Archimedes of Syracuse and Apollonius of Perga wrote principally for other mathematicians, they sometimes dedicated their treatises to notable heads of state, doing so in some cases not merely out of tact, but in a way that shows they expected their work to be understood. The Ptolemies tried to attract distinguished scientists to Alexandria for much the same reasons as they collected the texts of the masterpieces of Greek literature, namely to add lustre to their own reputation. In both cases other Hellenistic rulers pursued similar policies for similar motives—the Attalids too were avid collectors of manuscripts for their library at Pergamum, and museums were founded in many cities—although the Ptolemies were generally able to out-manœuvre or outbid their rivals.

The effects of this increasing kingly patronage, important though they are, should not be exaggerated. It would certainly be wrong to assume that every scientist who is recorded as having worked in Alexandria (and they include almost all of the most important names in third- and second-century science) was subsidized by the Ptolemies. Moreover the patronage of individual rulers was capricious. The first three Ptolemies were generous in their support of science, but later members of the dynasty were often far less so, and some (such as Ptolemy Physcon) even positively discouraged scientists from living in Alexandria. The Museum continued in existence right down to the fifth century A.D., but its fortunes, and those of the many lesser foundations modelled on it, fluctuated from one generation to another.

Institutions such as the Alexandrian Museum aided science both directly, through the financial help that scientists could obtain, and indirectly, by providing focal points round which groups of scientists gathered. But the economics of scientific research were scarcely less precarious in the Hellenistic than in the classical period. Even at Alexandria in the third century—the heyday of the patronage of science in the ancient world—the support given to science was modest in scale and uncertain in operation, as any support that ultimately depended on the generosity of an individual—the king himself—must be. There were many scientists who received no help whatsoever from rich patrons. Many of those who did scientific work were no doubt men of means. To give just one example, Archimedes evidently belonged to a wealthy and well-connected family: he is reported as being a friend and kinsman of Hiero, the king of Syracuse. Again many, perhaps most, of the scientists we shall be discussing in this study earned their living, or at least supplemented their means, either by teaching or by practising such professions as medicine and architecture. Most of the famous anatomists and physiologists in the ancient world were also, and often primarily, medical practitioners, and although the medical profession was highly competitive, successful physicians earned large sums of money both at Alexandria and later at Rome. There is less evidence concerning those who were employed as *architektones* (a term used not only of architects in our sense, but also of engineers and those in charge of the design and maintenance of weapons of war), but this too represented an important means of livelihood for some who engaged in scientific research. At no stage in the ancient world did science as such provide a livelihood in the sense that a person who wished to engage in scientific research might reasonably expect to make a living from that alone.

So far as the economic background of scientific research is concerned, the differences between the Hellenistic and earlier periods are a matter of degree, not of kind. Despite the changes I have mentioned, the continuity in the conditions under which scientific research was conducted before and after Aristotle is striking and this applies equally to what the ancient writers themselves say concerning their aims and motives. Here the picture that emerges is, to be sure, a complex one. Many writers refer to the ideal of the philosophical or contemplative life, a theme with many variations, in some of which research is valued for leading to a greater appreciation of the beauty of the universe or of the purposefulness of its creator, in others because it is claimed to improve men's characters. Other texts show that in some contexts scientific knowledge was valued not merely for its own sake, but also for its possible practical applications to the needs of mankind. Yet most of the main ideas that we find expressed on this question by scientists after Aristotle can be paralleled in the fifth and fourth centuries B.C. To put it another way, no *new* rationale or justification for scientific research was developed after Aristotle. Although scientists were helped by such institutions as the Alexandrian Museum or by enlightened—or simply ambitious—rich men, substantial and continuous state support for science was not forthcoming. In particular the idea that science holds the key to material progress, and indeed the ideal of material progress itself, were largely lacking.

The chief problem we must explore in this study is set by the circumstances I have just outlined. Given that science is a modern, not an ancient, category, the history of differing conceptions of the aims and justifications of different modes of the inquiry concerning nature will be one of our main concerns. In considering just a few of the individuals who embarked on that inquiry, we shall try both to illustrate briefly their specific problems, methods, arguments and theories, and to review such evidence as relates to their views of themselves and of the investigations on which they were engaged. I turn first to the immediate successors of Aristotle in the school he founded, the Lyceum.

2

The Lyceum after Aristotle

THE chief centre of research in mathematics, astronomy and biology in the third century B.C. was Alexandria, but Athens remained pre-eminent for 'philosophy', and this was generally defined to include not only logic and ethics but also 'physics', the study of nature and in particular the study of the ultimate constituents of matter. The most comprehensive physical system in the ancient world was that of Aristotle, who not only put forward detailed theories on such problems as the constitution of matter, but also laid down the types of question to be asked, and the terms in which to answer them, in his doctrine of the four 'causes', material, formal, final and efficient. Although his qualitative element theory, in which the four simple bodies, earth, water, air and fire, are analysed in terms of two pairs of primary opposites, hot and cold, and dry and wet,[1] did not lack competitors in the third century or later, this was, for long periods in antiquity, the dominant physical theory, as we can judge, for example, from its influence on Galen in the second century A.D. Our first task, in this chapter and the next, is to consider first the work of Aristotle's two immediate successors as heads of the Lyceum, namely Theophrastus and Strato, and then the rival physical systems proposed by other Hellenistic philosophers.

Theophrastus was born at Eresus in Lesbos, the son of a fuller, in about 371 B.C. and he was head of the Lyceum for some 36 years after Aristotle's death in 322. The range of his writings is almost as extensive as those of Aristotle himself. He wrote on logic, rhetoric, ethics and politics (for example *On Kingship* and *On the Best Constitution*), religion, metaphysics and numerous aspects of physics, as well as studies of his predecessors' ideas—in all over 200 separate treatises, many of which ran to several books. Apart from the famous *Characters*, the main works to have survived are the two great botanical

[1] See *Early Greek Science*, ch 8.

treatises, the *Inquiry Concerning Plants* in nine books and the *Causes of Plants* in six, but we also have a group of short, sometimes incomplete, physical treatises, an important essay on metaphysics and a portion of his history of psychology.

Theophrastus' debts to Aristotle are apparent in much of what he wrote. Yet he is also highly critical of Aristotle, both of his specific physical theories and of his general doctrine of causation, and it is indeed chiefly for these criticisms that Theophrastus interests us. His discussion of the doctrine of final causes—where Plato especially had foreshadowed Aristotle's view—is typical. While not rejecting this doctrine entirely, he devotes the last few pages of his essay on *Metaphysics* to questioning its range of application. His two main criticisms are first that to identify final causes is more difficult than had generally been acknowledged, and secondly that many things do not occur for the sake of an end at all. What purpose, he asks, do the tides serve, or breasts in male animals? Some things—his example is outsize horns in deer—are even harmful to the animals that possess them. 'We must', he says (11 a 1 ff), 'set certain limits to the final cause and to the tendency towards what is best, and not assume that it exists in absolutely every case . . . For even if this is the desire of nature, yet there is much that does not obey nor receive the good.'

The upshot of Theophrastus' discussion is certainly not to abolish final causes; indeed he criticizes thinkers who had *under*estimated their role. Yet he insists that to search for a final cause is in some cases mistaken. Although Aristotle himself had given a lead in this direction by contrasting what happens 'from necessity' with what happens 'for the sake of the better', Theophrastus spells out more clearly than Aristotle had done that 'there is much that does not obey nor receive the good'. He holds that the universe is in general well-ordered, and like most Greeks he believes that the heavenly bodies, in particular, manifest order to the highest degree. But he acknowledges and points out that in nature—both in animals and in the movements of the elements—a good deal happens at random and without an end.

A second example where Theophrastus subjects a key Aristotelian doctrine to a far-reaching critique concerns the theory of the simple bodies. In his short work *On Fire* he points out the differences between fire and the other simple bodies, air, water and earth. First, fire alone is able to generate itself: the other three simple bodies do not do so, although they change naturally into one another. Secondly, most of the ways in which fire comes to be, whether naturally or artificially, involve force. Thirdly, whereas we are not able to create the other simple bodies, we can make fire, and do so in many different ways. Fourthly, the greatest difference between it and the other simple bodies is that they are self-subsistent and do not require a substratum, whereas fire does. 'Everything that burns is always as it were in a process of coming-to-be and fire is a kind of movement; and it perishes as it comes to be, and as soon as what is combustible is lacking, it too itself perishes' (III 51 24 ff). Accordingly, 'it seems absurd to call this a primary element and as it were a principle, if it cannot exist without matter'.

Theophrastus is a penetrating critic of Aristotelian positions, but his criticisms rarely provide the basis of new solutions to the problems to which they relate. The *Metaphysics* raises doubts about the doctrine of final causes and exposes some of the weaknesses in such other Aristotelian theories as that of the first unmoved mover, but although it explores difficulties, it has little to offer in the way of positive new suggestions for their resolution. Similarly *On Fire* hits on a fundamental objection to Aristotle's doctrine on that subject. Yet Theophrastus attempts no new element theory, nor even a definite statement of what fire, in particular, is. He alludes briefly to the question of the nature of the primary pairs of opposites, remarking that on one view 'it appears that hot and cold are attributes of certain things, not principles . . . At the same time the nature of the so-called simple bodies is mixed and they exist in one another. Just as fire cannot exist without air or something wet and earthy, so the wet cannot exist without fire, nor earth without the wet' (III 53 3 ff). But although this leads him to say that the whole question of the definition of a simple body is one that needs further investigation, he draws back from that inquiry, partly, no doubt, because it is beyond the scope of his present problem, but partly also, one may presume, because of the difficulty of the question itself. The remainder of the treatise is devoted to a mass of major and minor problems, ranging from the various methods by which fire is made or extinguished, to explaining why, when we grip a hot spit tightly, we are burnt less severely than when we hold it loosely. His discussion contains many interesting observations, even though, in the absence of an objective temperature scale, Theophrastus naturally often relies too much on subjective impressions in judging hot and cold. Yet although he provides some important data relevant to the investigation of what fire is, he proposes no definite answer to the question of its nature or to the even larger problem of the ultimate constituents of physical bodies.

So far we have considered Theophrastus mainly as a critic of received opinions. The more positive side of his work is seen in such treaties as *On Stones* and the botanical works. Here he appears as a research worker in the tradition set by Aristotle in his zoology. The subject of *On Stones*, for instance, corresponds to our petrology. The framework of his theory, as so often, is derived from Plato and Aristotle, in that he classifies things found in the ground into two main kinds, those (like the metals) in which water predominates, and those (the 'stones' and the 'earths') in which earth does. But within this framework he attempts an account of the different kinds of stones, distinguishing them by such criteria as colour, hardness, smoothness, solidity, weight (that is what we should call specific gravity) and, perhaps most interestingly of all, by their reactions to other substances, especially fire and heat. In this last connection he remarks, for instance, that some metal-bearing ores melt and become fluid when subjected to fire; that some stones break and fly into pieces; that others burn (for example marble which, when burnt, produces quicklime) and that others again resist fire or are quite incombustible.

His discussion encompasses a remarkable range of data. Throughout the treatise he gives detailed information about the substances he describes, often specifying the precise locality from which a particular kind of stone, earth or metal comes. His use of data derived from industrial processes is especially interesting. Section 16, which describes the mining of a substance that is probably lignite, contains the first extant reference in a Greek text to the use of a mineral product as fuel. Sections 45–47 on the touchstone contain our first account of a method to determine the precise proportions of the constituents of an alloy, and among the many descriptions of the preparation of pigments is the first extant account of the manufacture of white lead. Among the processes he describes as having recently been discovered are the production of red ochre and the extraction of cinnabar. In section 60 he remarks that 'art imitates nature and also creates its own peculiar substances, some of them for use, some merely for their appearance . . . and some for both purposes, like quicksilver', and after stating briefly how this is produced, he adds: 'perhaps several other such things could be discovered'.

There is no evidence that Theophrastus tried to follow up his attempts to distinguish different kinds of stones and earths by conducting systematic experiments, even though some aspects of his inquiry might have suggested to him the value of doing so. On the other hand he collects a good deal of information—no doubt from a variety of sources, including earlier writers, as well as from his personal inquiries and observations—and evaluates it, in general, with care. Whereas most ancient writers on minerals deal at length with their magical properties, especially their supposed therapeutic powers, *On Stones* is almost entirely free from such ideas. True, he does, on some three or four occasions, apparently endorse a popular fable, as for example the story of the stone called *lyngourion* that was supposed to be produced—as its Greek name suggested—by the urine of a lynx. But he reports few such tall stories, and he expresses his doubts about most of those he does retail. As the chief modern editors of the work, E. R. Caley and J. F. C. Richards, put it: 'for almost two thousand years, this treatise by Theophrastus remained the most rational and systematic attempt at a study of mineral substances'.[1]

Finally his caution as a theorist, his meticulousness in marshalling data and his own skill as an observer are seen again in his botany. The *Inquiry Concerning Plants* and the *Causes of Plants* are his acknowledged masterpieces. They were evidently planned on the general model of Aristotle's zoological treatises, and like them they contain both minute descriptions of particular species and highly theoretical discussions of the

[1] *Theophrastus On Stones*, ed. E. R. Caley and J. F. C. Richards (The Ohio State University, Columbus, Ohio, 1956), p 10.

causes of the phenomena described. Most of the data they present were, no doubt, already known to nurserymen and farmers. Here Theophrastus' role is one of collecting, collating and organizing this knowledge. It is not, however, his aim to give a comprehensive classification of plants. On the contrary, he proceeds with typical caution on this subject. Although he identifies four main classes of plants ('trees', 'shrubs', 'under-shrubs'—*phrȳgana*—and 'herbs') he insists that the definitions he gives of these 'must be taken . . . as applying generally and on the whole. For in the case of some plants it might seem that our definitions overlap; and some under cultivation appear to become different and depart from their essential nature'[1] (*Inquiry* I 3 2). And again he notes (*Inquiry* I 4 3) that 'nature does not thus go by any hard and fast law [literally: possess necessity]. Our distinctions therefore and the study of plants in general must be understood accordingly.'

Much of both the *Inquiry* and the *Causes* is devoted to detailed accounts of particular species of plants. But to give some idea of his approach to the problems that his study raises we may take just one example, his discussion of spontaneous generation. His general view concerning the different modes of reproduction of plants is given in the *Inquiry* II 1 1, for example:

> The ways in which trees and plants in general originate are these: spontaneous growth, growth from seed, from a root, from a piece torn off, from a branch or twig, from the trunk itself. . . . Of these methods spontaneous growth comes first, one may say, but growth from seed or root would seem most natural; indeed these methods too may be called spontaneous; wherefore they are found even in wild kinds, while the remaining methods depend on human skill or at least on human choice.[2]

As this passage shows, the term spontaneous (*automaton*) is used in two different but related ways, (1) to distinguish between the wild and the cultivated—where what happens 'of its own accord' is opposed to what happens by human agency;

[1] From the Loeb translation of Sir Arthur Hort (Cambridge, Mass., Harvard University Press; London, Heinemann, 1916).

[2] From the Loeb translation of Sir Arthur Hort.

and (2) to distinguish reproduction without seed from reproduction by seed, root or any other natural method. The problem of spontaneous generation in the second sense is discussed on several occasions. Greek writers had generally assumed that both animals and plants were spontaneously generated in certain substances under certain conditions. In the *Causes of Plants* (I 5 1 ff) Theophrastus begins:

> Spontaneous generation, broadly speaking, takes place in smaller plants, especially in those that are annuals and herbaceous. But still it occasionally occurs too in larger plants whenever there is rainy weather or some peculiar condition of air or soil . . . Many believe that animals also come into being in the same way.[1]

Yet having thus apparently assented to the common view, he goes on to introduce reservations:

> But if, in truth, the air also supplies seeds, picking them up and carrying them about, as Anaxagoras says, then this fact is much more likely to be the explanation . . . Moreover the rivers and the gathering together and breaking forth abroad of waters purvey seed from everywhere . . . Such growths would not appear spontaneous, but rather as though sown or planted. Of the sterile sorts, one might rather expect them to be spontaneous, as they are neither planted nor grown from seed, and if they come to be in neither way, they must necessarily be spontaneous. But this may possibly not be true, at least for the larger plants; it may be rather that all the stages of development of their seeds escape our observation, just as was said in the *Inquiry* about willow and elm.

Although he sees very clearly that many cases of what was believed to be spontaneous generation are not such, Theophrastus does not reject the notion of spontaneous generation itself. The chapter concludes:

> But when this phenomenon is found in fruit-bearing and flowering plants, what prevents it occurring in others not fruit-bearing? Let this be given merely as our opinion; more

[1] Based on the translation of R. E. Dengler, *Theophrastus De Causis Plantarum Book One* (Philadelphia, 1927).

> accurate investigation must be made of the subject and the matter of spontaneous generation must be thoroughly inquired into. To sum the matter up generally: this phenomenon is bound to occur when the earth is thoroughly warmed and when the collected mixture is changed by the sun, as we see also in the case of animals.

As, in the *Metaphysics*, his attack on the doctrine of final causes was a critique of applications of the idea, not of the idea itself, so, in his botany, he raises objections to many currently accepted examples of spontaneous generation, but does not deny the doctrine as a whole. He demands further inquiry, but this is into particular cases: he does not doubt that spontaneous generation sometimes takes place. We should, however, note that, assuming that life has not always existed on earth, it was, *prima facie*, no less reasonable to believe that living things can *continue* to be generated from non-living matter under certain conditions as to hold that life emerged from the non-living at one particular moment in the past. Theophrastus accepts spontaneous generation, while acknowledging it to be problematic, and he attempts to explain it in naturalistic terms as due to the action of the sun on the moist earth.

Theophrastus was succeeded by Strato of Lampsacus, who was head of the Lyceum from about 286 to 268. He too evidently wrote on a wide variety of subjects, including logic, ethics and politics. But as his nickname, *ho Physikos*, suggests, his main interest was the study of nature, especially those branches of it that we should call physics and dynamics, although he also wrote on zoology, pathology, psychology and technology. Unfortunately not a single one of his works has survived in its entirety and we have to piece together his ideas from the accounts and quotations in later writers.

His most interesting work concerns two problems in physics, namely gravity and the void. On the question of gravity—or, as the Greeks expressed it, the nature of the heavy and the light—Strato rejected Aristotle's doctrine that there are *two* natural tendencies, one of heavy bodies towards the centre of the earth and the other of light ones away from it. He saw that there was no need to postulate a separate upward tendency to account for the movement of air and fire, for example, since this can be explained as being due to their being displaced by the downward movement of heavier bodies.

Thus far all that Strato had done was to replace one theory with another, rather simpler one. But these reflections on the nature of weight were followed up by an inquiry concerning acceleration. A passage from his treatise *On Motion* is quoted by Simplicius in his *Commentary on Aristotle's Physics* (916 10 ff). Simplicius notes that different explanations of acceleration are given by different writers, but adds that 'few adduce any proof of the fact itself, that when bodies moving naturally are near their natural places they move more swiftly'. Strato, however, did so:

> For in his treatise *On Motion*, after asserting that a body so moving completes the last stage of its trajectory in the shortest time, he adds: 'in the case of bodies moving through the air under the influence of their weight this is clearly what happens. For if one observes water pouring down from a roof and falling from a considerable height, the flow at the top is seen to be continuous, but the water at the bottom falls to the ground in discontinuous parts. This would never happen unless the water traversed each successive space more swiftly' . . .
>
> Strato also adduces another argument, as follows: 'if one drops a stone or any other weight from a height of about a finger's breadth, the impact made on the ground will not be perceptible, but if one drops the object from a height of a hundred feet or more, the impact on the ground will be a powerful one. Now', he says, 'there is no other cause for this powerful impact. For the weight of the object does not increase, the object itself has not become greater, it does not strike a greater space of ground, nor is it impelled by a greater [external force]: rather it moves more quickly.'[1]

The importance of this text lies not so much in what Strato was trying to prove (the fact that acceleration occurs) as in the way he tried to prove it. First he appeals to a phenomenon that anyone could observe, rain-water falling from a roof.

[1] Based on the translation in M. R. Cohen and I. E. Drabkin, *A Source Book in Greek Science* (second edition), Cambridge, Mass., Harvard University Press, 1958.

Secondly he adduces the evidence of what would happen if a given weight were dropped from different heights. The imprecise way in which this test is described makes it likely that it is no more than a 'thought experiment', but we have other evidence that suggests that in other cases Strato actually carried out experiments on physical problems.

Experiment is a feature of the method he used on the second topic I mentioned, the nature of the void. Here a passage in the introduction to the *Pneumatics* of Hero of Alexandria (first century A.D.) provides our chief testimony. This contains embedded in it a text which we know on independent evidence to be a verbal quotation from Strato, and as elsewhere too the theories correspond to his, it has been thought likely that much of this introductory passage in Hero derives from him, although it certainly contains material from other sources as well.

One section of the introduction that probably derives from Strato is the description of a simple test to show the corporeality of air, in which an empty vessel is inverted and pressed down into water. Here 'the air, being a body, does not allow the water to enter', although (as the passage continues) if a hole is bored in the bottom of the vessel and the test repeated, the water will enter as the air escapes through the hole.

A more striking series of experiments, this time designed to prove that a continuous vacuum can be produced artificially, follows a little later in Hero. This passage first attacks those (like Aristotle) who denied absolutely that a void can exist, accusing them of putting their faith in argument rather than sensible evidence:

> Those then who assert generally that there is no vacuum are satisfied with inventing many arguments for this and perhaps seeming plausible with their theory in the absence of sensible proof. If, however, *by referring to the appearances and to what is accessible to sensation*, it is shown that there is a continuous vacuum, but only one produced contrary to nature; that there is a natural vacuum, but one scattered in tiny quantities; and that bodies fill up these scattered vacua by compression; then those who put forward plausible arguments on these matters will no longer have any loophole. (I 16 16 ff).

After this statement of the superiority of sensible evidence to abstract argument, Hero proceeds to describe the apparatus:

> Prepare a sphere, of the thickness of metal plate so as not to be easily crushed, containing about 8 cotylae ($\simeq$ two quarts). When this has been tightly sealed on every side, pierce a hole in it and let in a siphon, or thin tube of bronze, so that it does not touch the part diametrically opposite the hole but allows a passage for water, and so that the other end of the siphon projects about three fingers' breadth above the sphere. The circumference of the hole through which the siphon is inserted must be sealed with tin applied both to the siphon and to the outer surface of the sphere, so that when we breathe through the siphon, the air cannot possibly escape from the sphere. Let us observe what happens.

The first experiment is designed to show that there are scattered vacua in air. If there were no such vacua, it was argued, then air should not be compressible. 'And yet if you insert the siphon in your mouth and blow into the sphere, you will introduce much breath in addition, without the air that is contained in the sphere escaping. And since this always happens, it clearly shows that a compression of the bodies in the sphere takes place into the interspersed vacua.' Next it is shown that air can be evacuated from the globe: 'again if you want to draw out, with your mouth, through the siphon, the air that exists in the sphere, a fair quantity will come out, although no other substance takes its place in the sphere . . . So that by this means it is conclusively proved that a considerable accumulation of vacuum occurs in the sphere.'

The experiments described in this passage have quite probably been elaborated either by Hero himself or by other writers intermediate between him and Strato. Yet we may be fairly confident that some such set of tests goes back to Strato himself, for the text continues, after a digression, with what we know to be in part a direct quotation from him. Here a further argument, based on the penetration of heat and light through water and air, is brought to show that scattered vacua exist. Both in his study of falling bodies, and in his investigation of the void, Strato's method seems to have consisted in appealing to the twin evidences of nature and of art—

that is, of natural phenomena observed directly, and of deliberate experiments.

Poor though our evidence for Strato is, it nevertheless suggests that, more than any earlier Greek scientist, he tried to use experimentation to investigate physical problems, and although most of his actual experiments are inconclusive, this seems to us the most important feature of his work. The extent of his influence on later scientific method is, however, very uncertain. Later writers in the inquiry known as 'pneumatics', such as Philo and Hero, certainly appear to draw heavily on him. Yet whereas the tests that Strato himself conducted were intended to demonstrate certain propositions, such as that air is a body, the motive of later writers on pneumatics in performing similar experiments is often very different. They were interested less in proving or disproving particular theories than in the phenomena that could be produced artificially in themselves. They sought to create striking or curious effects for their own sakes. There is a strong tradition of experiment in this loose sense—the trying out of new effects—that we shall consider further when we discuss the relations between science and technology in Chapter 7. Yet the purpose of those investigations is a practical one, and they were not designed to demonstrate or refute physical theories.

Both Theophrastus and Strato rejected many of Aristotle's ideas, yet both obviously remained profoundly influenced by him. Theophrastus criticized the doctrine of final causes, but remained a teleologist nevertheless. In general Strato's position is further from Aristotle's: thus he seems to have denied the operation of final causes in nature as a whole, for, according to some reports, he both rejected the doctrine that the world is a product of divine providence, and reduced nature to 'chance' and 'the spontaneous'. And he broke new ground in his attempts to investigate problems in dynamics and pneumatics experimentally. Yet his element theory is eclectic: like the atomists, among others, he appears to have assumed that the underlying structure of matter is particulate, but he also held that the fundamental principles are the two primary qualities, hot and cold, a doctrine that clearly owes most to Aristotle. Both Theophrastus and Strato did original work in particular branches of natural science. But neither produced a comprehensive physical theory to rival Aristotle's. Yet such theories were put forward, in the generation after Aristotle, as parts of Epicureanism and Stoicism, and our next task is to consider these, the two most important new philosophical systems of the Hellenistic age.

3

Epicureans and Stoics

Both Epicureans and Stoics divided philosophy into three parts, ethics, physics and logic, and both subordinated physics and logic to ethics. Both schools insisted that the chief aim of philosophy is to secure happiness, and that for a man to be happy he must be free from anxiety and fear. Since he will continue to be troubled by irrational fears so long as he is ignorant of the causes of natural phenomena, it follows that he must study physics as well as moral philosophy. Although the Epicureans and Stoics disagreed both on the question of the chief good and on many fundamental problems in physics, they both maintained that the main motive for investigating the latter is to obtain peace of mind.

Epicurus, who was born in Samos of Athenian parents in about 341 B.C. and who set up his school, the Garden, in Athens towards the end of that century, says explicitly in the *Principal Doctrines* (11), that 'if we were not troubled at all by apprehensions about phenomena in the sky and concerning death, lest it somehow concern us, and again by our failure to perceive the limits of pains and desires, we should have no need of the study of nature'. And in the *Letter to Pythocles* (85) he says, 'bear in mind that there is no other end to the knowledge of things in the sky . . . than peace of mind and firm conviction'. Superstition and mythology are repeatedly attacked. 'Solstices, eclipses, risings and settings' and so on take place 'without the ministration or ordering' of gods, and the regularity of phenomena in the sky is due to the arrangement of atoms, not to god (*Letter to Herodotus*, 76 f). But although natural science is used to counter false religious beliefs, there is no need to study physics beyond the point where one has reached some explanation for phenomena that might be thought to be due to supernatural intervention or that might otherwise give rise to irrational fears.

Epicurus' investigations of certain physical problems are, however, carried a good deal further than one might expect in view of the reasons for which they were undertaken. Our chief sources for his physics are the *Letters to Herodotus* and *to Pythocles* though the Latin poem *On the Nature of Things* written by his great follower Lucretius (first century B.C.) is also valuable. Although Epicurus himself denied any debt to earlier thinkers, his main physical doctrines are derived from the fifth-century atomists, Leucippus and Democritus. Like them, he held that atoms and the void alone exist. There is an infinite number of atoms and they are in constant movement in an infinite void. They collide with, and rebound from, one another, and so form complex bodies, but the sensible qualities that these bodies possess are not real but mere appearances. All such qualities as heat, colour, taste and so on are derived from, and reducible to, differences in the primary properties of atoms, such as shape and position.

Thus far Epicurus simply followed Democritus. But in certain parts of his theory he diverged from what Democritus had taught—in most cases, it seems, as a response to the criticisms that Aristotle had brought against atomism in its original form. One of the issues that the earlier atomists had left unresolved concerns the nature of the atoms themselves. Although they certainly held that the atoms are physically indivisible (that is, they cannot be split), it is not clear whether they believed that the atoms are mathematically divisible (that is, divisible at least in thought).[1] Thus Aristotle objected to the atomists on grounds which show that he believed that they failed to distinguish between physical and mathematical indivisibility. Epicurus' answer was to postulate two types of minima and to distinguish clearly between them. The atoms are physical minima, the unsplittable units of which physical objects are composed, but the atom itself has size and it contains, and is made up of, mathematically indivisible parts.

Again on the question of the shapes of the atoms, the earlier atomists may have assumed that just as the atoms are infinite in number, so too their shapes are infinitely various. But then it could be objected that if the shapes vary infinitely, then some of the atoms will be extremely large, certainly large enough

[1] See *Early Greek Science*, p 47.

to be visible. This was clearly unacceptable, and Epicurus' response was to argue that the shapes—and sizes—of the atoms are not infinitely various, only indefinitely so.

But the most important revision that he introduced concerns weight. Leucippus and Democritus held that the primary properties of atoms are shape, arrangement and position alone. To these Epicurus added weight—which the earlier atomists treated as a secondary property acquired when the atoms have collected to form a world. The process by which worlds are formed was, accordingly, interpreted quite differently by the earlier atomists and by Epicurus. Leucippus and Democritus held that the atoms are in eternal movement in all directions and that their chance collisions cause agglomerations which in turn attract other atoms. Epicurus, by contrast, imagined that before the formation of a world all the atoms are travelling in the same direction, that is 'downwards', in space. Moreover whereas Aristotle, for instance, had held that the heavier the body the faster it falls, Epicurus argued that in the void speed does not vary with weight, and that heavy and light atoms all fall 'as quick as thought'. But then no atom will meet or collide with any other—and no world will be formed—and so Epicurus put forward his famous, or rather notorious, doctrine of the swerve. Occasionally an atom deviates from the vertical by the smallest possible amount. There is no cause for this deviation—it is simply an effect without a cause. Yet such deviations must, he believed, occur, both at the beginning of the formation of a world and within worlds once they had been formed. Although the evidence is incomplete and in parts unclear, Epicurus apparently applied the doctrine of the swerve to his account of the soul, to rescue his moral philosophy from determinism.

Already in antiquity this doctrine was the subject of ridicule. Yet the moral argument is not so crass as has often been made out. As a materialist, Epicurus explained mental events in terms of the physical interactions of soul-atoms, and his problem was to say what moral responsibility could mean in such a system. It has commonly been supposed that his solution was to postulate a swerve in the soul-atoms for every 'free' action. Yet there is no direct evidence that this was his view, and indeed to account for *choice* by assuming the intervention of an *uncaused* event at the moment of decision is bizarre. It is more likely that (as has recently been argued[1]) Epicurus' account of responsibility, like Aristotle's, depends rather on his notion of character, and that the function of the swerve is merely to introduce a discontinuity at *some* point in the motions of the soul-atoms in order to make room for the possibility of free choice. The swerve need not, indeed should not, take place at the moment of choice: all that is necessary is that a swerve occurs at some stage in the soul-atoms to provide an exception to the rule that their interactions are fully determined.

The interpretation of the cosmological argument for the swerve is less controversial. Once it was assumed that all atoms move at the same speed in the same direction, then there must be exceptions to this if we are to explain the cosmogonical process. We know that this world, at least, exists: somehow or other, then, collisions of atoms must have occurred. Moreover his suggestion that a single atom deviates from time to time from the norm may be less shocking to us than it was to some ancients. The usual, though not the only, ancient conception of a physical law was of a necessity applying to every instance of a given phenomenon without exception—in contrast to the modern notion of a physical law as a statement of a statistical probability. Again, according to Heisenberg's Uncertainty Principle, the behaviour of a single nuclear particle cannot be completely described—that is, we cannot determine at the same time *both* its location *and* its momentum. Yet this superficial similarity should not obscure the fundamental differences between the motives and contexts of Epicurus' doctrine on the one hand, and those of nuclear physics on the other. When Epicurus postulated his exceptions to the rule that he thought was otherwise observed, he conceived that rule in terms of the usual ancient notion of an iron necessity—the very idea that the modern conception of a physical law has supplanted. His hypothesis of the swerve was not the result of an inquiry (whether logical or empirical) into the nature of the information that is available concerning the atoms, so much as a desperate measure to save at one stroke both his cosmology

[1] See D. J. Furley, *Two Studies in the Greek Atomists*, Princeton University Press, Princeton, 1967.

and his moral philosophy from the consequences of his physical theory.

Epicurus maintains that there is one account, and one account only, to be given concerning such fundamental principles as that nothing exists besides atoms and the void. But when it comes to giving the causes of particular natural phenomena, he applies a different method of explanation. Here he insists that there may be, and usually is, more than one cause for any given phenomenon. If several possible explanations suggest themselves, then so long as none of them is positively contradicted by the evidence, all should be allowed to stand. In the *Letter to Pythocles* (86 f) he says that celestial phenomena 'admit of manifold causes': 'if we accept one of these and reject another that is equally consistent with the appearance, we evidently leave the study of nature completely and rush headlong into myth'.

In practice he applies this principle not only to such questions as the nature of thunder and lightning or comets or hail or whirlwinds—where the state of knowledge at the time was primitive—but also to astronomical problems on which much successful research had already been done. Here, for example, are his remarks on the phases of the moon from the *Letter to Pythocles* (94):

> The wanings of the moon and again its waxings could come about through the revolution of its body, or equally well through the configurations the air assumes, or again through the interposition of other bodies, or in any other of the ways in which phenomena on earth invite us to explain such an appearance, provided that one is not so much in love with the method of a single explanation that one groundlessly rejects the others without having considered what it is possible—and what impossible—for a man to investigate.

In this, and many similar passages, the unscientific, indeed anti-scientific, aspects of Epicureanism become evident. He believes that research is futile if it does not contribute to peace of mind. He rules out further inquiry into which of several conflicting explanations of a phenomenon is correct and abuses those who attempted such investigations first for being dogmatic on topics where the principle of plural causes applies, and secondly for indulging in superstition and mythology. Yet those very astronomers whom Epicurus dismisses with contempt had long ago given the correct explanation of the phases of the moon.

Epicureanism is a strange amalgam in which serious inquiry into some of the basic problems of physics is combined with an essentially negative attitude towards research into particular natural phenomena. In introducing certain modifications into atomic theory in answer to its critics, Epicurus made a contribution to the debate concerning the ultimate constituents of matter. But on detailed issues in such fields as astronomy, once the investigation had yielded some plausible explanation or explanations and the initial temptation to believe the myths had been removed, then inquiry ceased. Epicureanism provides a clear illustration of how underlying philosophical assumptions influenced the nature of scientific work, but in this case the influence of those assumptions was two-edged, for they inhibited empirical research more powerfully than they encouraged abstract investigation into fundamental physical problems.

The rival, and in the long run more influential, philosophical system of Stoicism was founded and developed by three men, Zeno of Citium, Cleanthes of Assus and Chrysippus of Soli. All three worked chiefly in Athens, the first two being slightly younger contemporaries of Epicurus, the third some fifty years their junior. Whereas the doctrines of the Epicureans remained remarkably constant throughout the long life of the school—as we can see from Lucretius—Stoicism underwent many modifications, not only in its earliest period, but also and more especially in the periods of the so-called Middle and Late or New Stoa, at the hands of such men as Panaetius of Rhodes (born c. 185 B.C.), Posidonius of Apamea (c. 130–50 B.C.) and Seneca (first century A.D.). We are concerned here, however, with the doctrines of the founders of the school. The principal physical theories were mostly the work of Zeno himself, while Chrysippus was responsible for many of the detailed doctrines and arguments that helped to consolidate the system against the criticisms of its opponents.

The underlying motive for the study of natural phenomena was, as we have said, the same for the Stoics as for the Epicureans, namely the achievement of peace of mind. Otherwise, however, the two schools were in fundamental disagreement on almost every important issue in physics. Whereas Epicurus held that atoms and the void alone exist, the Stoics denied that there is a void within the world, although outside the world there is infinite void. They rejected the argument that the void is necessary to account for movement. The world is a plenum, but that does not prevent movement taking place within it. As fish move through water, so any object can move through the plenum, this being conceived as an elastic medium. Against the view that matter exists in the form of indivisible units, the Stoics held that the world is a continuum whose substance is infinitely divisible. Again whereas Epicurus treated space and time as composed of minimal parts, the Stoics conceived them as continua.

In contrast to the essentially quantitative theory of the atomists, in which qualitative differences are reduced to differences in the shape, arrangement and position of the atoms, Stoic physics was essentially qualitative. The starting-point of their cosmology is the distinction between two fundamental principles, the active and the passive. The passive is qualityless substance or matter. The active is variously identified as the cause, god, reason, *pneuma* (breath or the vital spirit), soul and fate. *Both* these principles are corporeal, and to describe the relation between them the Stoics used the term *krāsis di' holōn*, 'total mixture', putting to service an original theory of different modes of composition. *Krāsis* is defined as the total interpenetration of two or more substances, as exemplified in the mixture of two liquids such as wine and water. It is distinct both from *parathesis*—the mere juxtaposition of parts, as in a mixture of two kinds of seeds—and from *synchysis*—where, as in what we should call chemical combination, the component substances are destroyed and a new substance is formed as a result. The active principle is, then, thought of as inherent in the whole world and permeating every part of it.

These two fundamental principles are ungenerated and indestructible. But the physical elements of material bodies, like the cosmos itself, are generated and destroyed. Following

Empedocles, Plato and Aristotle, the Stoics held that all other physical substances are composed of four elements: fire and air, associated with hot and cold respectively, are the more active elements, water and earth, associated with wet and dry respectively, the more passive ones. The process of cosmogony begins when fire changes to first air, then water, then earth. Conversely the world is periodically destroyed when the process is reversed and the other elements are changed back into fire. The world thus begins and ends in fire in a process that repeats itself innumerable times.

Thus far the Stoic element theory and cosmogony stayed close to traditional forms, but in their doctrine of *pneuma*, breath or spirit, they were more original, though this idea too owes something to earlier philosophy. To judge from our conflicting testimonies, the interpretation of this doctrine already caused the ancients some difficulty. But despite one text that implies that *pneuma* is a fifth element on a par with the other four (like *aithēr* in Aristotle[1]), it is fairly clearly to be identified with, or seen as a form of, the active principle: and since this is present in everything, so *pneuma* must pervade all things, as indeed several authorities state. *Pneuma* is said to consist of air and fire in so far as these are the more active elements. According to Galen (K VII 525 11 ff), 'the *pneuma*-like substance causes cohesion, the matter-like substance is made cohesive, and so they say that air and fire make cohesive and that earth and water are made cohesive'.

The doctrine of *pneuma* leads us to that of the different levels of unified structure. Unlike discrete wholes, such as an army, or contiguous entities, composed of conjoined elements, such as a ship, where any of the elements can continue to exist even though all the others are destroyed, a unified structure is one where there is *sympatheia*, common affection or affinity, between the parts, so that whatever affects a single part affects the whole. But unified structures are of different kinds. First, the simplest type, seen in such things as stone, wood and metal, has what is termed *hexis*, a state of being held together. Secondly, at a more complex level of organization, there is nature, *physis*, the type of unified structure exhibited for example by

[1] See *Early Greek Science*, pp 109 ff.

plants. Thirdly, animals have not only *physis*, nature, in so far as they grow and reproduce themselves, but also *psȳchē*, soul, in so far as they are capable of movement and sensation. Unified entities are, then, distinguished according to the type of structure they possess, but in each case what holds them together is *pneuma*, conceived not as a static, external, constraining force, but as a dynamic, internal tension.

This doctrine is applied to the universe as a whole, for this too is a 'unified' entity in the sense defined. Chrysippus, for example, is reported as believing that the entire substance of the world is unified by a *pneuma* which wholly pervades it. The whole is a dynamic continuum in which, as in a living organism, all the parts are in a state of tension and transmit to one another whatever affects them.

The analogy between microcosm and macrocosm was a favourite one with the Stoics, being, indeed, no mere analogy. The universe is not simply *like* a living creature, it *is* one. Like a human being, it is pervaded with vital spirit (*pneuma*), life/soul (*psȳchē*) and reason (*nous*). The active principle is not only a principle of cohesiveness, but also a generating force. In accounts of the cosmogonical process god is described as the 'seminal formula' of the universe, and his generative activity is conceived as like that of animal seed. The reason that pervades the universe is, moreover, not merely an intelligent, but also a providential, cause. But divine providence implies no break in the chain of physical causes and effects. On the contrary, it is, rather, the *name for* that chain: the sequence of cause and effect is both fate and the will of god. Chance is eliminated, or rather it is interpreted purely subjectively as that which is obscure to human cognition. Although in the ethical sphere the Stoics did not deny moral responsibility—arguing, with the help of a distinction between different kinds of cause, that some part of our actions is 'in our power' in the sense of depending on our character—they were consistent physical determinists. Once again the contrast with Epicurus is striking. Whereas he denied that the world is a living creature, that it exhibits design, and that there is an inexorable chain of cause and effect, the Stoics maintained all three doctrines. Moreover they held not only that the future is determined, but also that it can, in principle, be predicted, and this they attempted to do by practising various kinds of divination.

We cannot do justice here either to detailed doctrines or to controversial points of interpretation, but enough has perhaps been said to give the main outlines of early Stoic physics. This is the first fully elaborated continuum theory of matter in antiquity. Other philosophers had denied the existence of a void and had attempted an analysis of different modes of mixture. But the Stoics were the first to work out a detailed physical system based on the notion of a continuum in which all the parts intercommunicate. The conception of *pneuma*, the theory of total mixture or interpenetration, *krāsis*, the doctrine of *sympatheia*, the conception of space and time as continua, all form part of a carefully elaborated and remarkably consistent whole. Their conception of the chain of cause and effect followed from this theory. Their belief in the possibility of predicting the future is, in turn, consistent with, and indeed entailed by, their determinism. To label their attempts at divination 'unscientific' is beside the point. We must recognize that there was, in their view, no difference in kind between the physicist's attempt to arrive at general laws by induction and the diviner's attempt to foretell the future by using his observations and his 'art'. The comparative success of their predictions might vary, but the rationality of their procedures did not. In the Stoics' eyes, the rational basis for the practice of divination, as for science itself, is the philosophical belief in the unbreakable chain of cause and effect.

The physics of Epicurus and the early Stoics were abstract systems based on reflection on the problem of the ultimate constitution of matter. Although both schools proposed causal explanations of many astronomical, meteorological and biological phenomena, neither engaged in empirical research to any significant extent. Nevertheless their debate brought into the open the issue between two fundamentally opposed, but complementary, conceptions of matter—atomism and continuum theory. Either matter exists in the form of discrete particles, separated by void, or it is a continuum of intercommunicating parts. With each conception is associated a different doctrine of movement—a particle, and a wave, theory: either motion is the transport of material particles through the void, or it is the transmission of a disturbance in an elastic medium. Despite the fundamental differences between different theories that have shared the same name, Greek atomism may be considered the prototype of all theories in which matter is conceived as particulate in form, and Stoic physics, in turn, may be seen as the forerunner of later continuum theories of matter. Indeed the chief modern authority on Stoic physics, S. Sambursky, has suggested that we can see in the doctrine of *pneuma* interpenetrating matter 'the prototype of the field of force as it was developed in the physics of the nineteenth century', even though he adds: 'the significant difference being that in the age of mathematical physics this latter field concept was entirely stripped of any substantiality in the purely material sense of the word.'[1]

Natural science is intimately linked with ethics and with theology in both Epicureanism and Stoicism. Both schools connected the problem of causation in general with the moral question of choice and responsibility. Both schools also saw an understanding of the true nature of the gods as one of the benefits accruing from the study of physics. Most importantly, the motive for the inquiry concerning nature is, in each case, an ethical one, and this influenced fundamentally both the types of subject they investigated and the way they investigated them. Whereas both schools held that some knowledge of the basic problems of physics was essential for peace of mind, neither provided any incentive to examine any of the special branches of science in any detail. Rather their emphasis on the practical moral aims of speculative inquiry—the view that science is a means to an end, not an end in itself—acted as a strong disincentive to their undertaking any such study. Despite the importance of their work on the general theory of matter, neither Epicurus nor the early Stoics[2] made any significant

[1] *Physics of the Stoics*, Routledge and Kegan Paul, London 1959, p 37.

[2] Among later Stoics, however, there are exceptions. The most important of these is Posidonius, whose work cannot be discussed here for reasons of space beyond noting that, although most aspects of the assessment of this highly controversial figure remain disputed, the evidence that he undertook empirical researches on certain physical problems is good. Thus we know from Strabo, who composed his *Geography* at the beginning of the Christian era, that Posidonius carried

contribution to any field of science that depended largely on observation. The great advances made in such subjects as astronomy and biology in the third and second centuries B.C. were the work of men who, while often influenced by philosophical assumptions, were not themselves primarily philosophers, but either mathematicians or doctors, and it is to their work that we must now turn, beginning with a brief survey of mathematics itself.

---

out observations of the periodicities of the tides, of which he was the first to offer a satisfactory comprehensive explanation.

4
Hellenistic Mathematics

So far we have been dealing with the work of men who thought of themselves primarily as philosophers. We must now consider a group who show little or no interest in the traditional problems of philosophy (whether physics, ethics or logic) and most of whom would have disclaimed the title of philosopher in preference for that of mathematician. The term *ta mathēmata*, derived from *manthanein*, to learn, is applied in Greek not only to what we should call mathematical studies but also generally to any branch of learning. Thus in the fourth century Plato used it of the study of the Form of the Good as well as of arithmetic or calculation, plane and solid geometry and astronomy in the *Republic*. Aristotle was the first person to distinguish systematically between *mathēmatikē* and *physikē*. The latter studies natural objects as such, nature being defined in terms of a capacity for movement or change. The objects of the former (planes, lines, points and numbers), although inseparable in fact from sensible objects, are studied in abstraction from them. Within the branches of mathematics he recognized a hierarchy. The primary studies are arithmetic, plane and solid geometry. Subordinate to these are the 'more physical' branches of mathematics, optics, harmonics, mechanics and astronomy, within which—modifying a thesis argued emphatically by Plato—he drew a further distinction between, for example, mathematical optics (a special application of geometry) and physical optics: this last includes, for instance, the study of the rainbow and is considered a special application of mathematical optics.

The turning point in the development of mathematics as a deductive inquiry came during the fifth century B.C. Long before that—and long before the Greeks engaged in this study—many mathematical propositions had been discovered, and many mathematical operations had been perfected, by the Egyptians and Babylonians. But the attempt to produce rigorous demonstrations of mathematical theorems was a new departure, for which Greek mathematicians of the fifth century were largely responsible. Very little of their work is, however, extant, and we have to rely, in almost every case, on the reports and allusions in later writers. From the end of the fourth century our sources improve dramatically. The first major mathematical text to have come down to us is the *Elements* of Euclid. In addition to other books of his that have survived, a substantial portion of the work of both Archimedes and Apollonius is also extant, providing rich material for the study of the development of mathematics in the 150 years after the death of Aristotle. In commenting on this I shall concentrate on the evidence that throws light on the aims and assumptions of mathematics, its methods and its applications to other fields of science.

We know little about Euclid the man. Most of our information comes from a brief account in the most important of his many Greek commentators, Proclus (fifth century A.D.), who tells us that he lived 'in the time of the first Ptolemy' (died 283 B.C.) and that he was 'younger than the pupils of Plato but older than Eratosthenes and Archimedes'.[1] It is clear from the phrasing of this report that Proclus himself had no precise information concerning the date and place of Euclid's birth. But the view that he was active around 300 B.C. is very probably correct. Whether or not he was born in Alexandria, several of our sources connect him with that city. Proclus tells the story of his reply to Ptolemy that 'there is no royal road to geometry', when Ptolemy had asked whether there was any shorter way in the subject than that of the *Elements*. And we may infer that Euclid taught at Alexandria from a passage in Pappus (*Mathematical Collection* VII, 35, 678 10 ff) that reports that Apollonius spent a long time there with Euclid's pupils. Besides the *Elements*, he also wrote on astronomy, optics and musical theory, and several of his shorter works—for example the *Optics*—are extant, although some of the books ascribed to him, such as the *Catoptrics* or *Theory of Mirrors*, are not authentic.

The relation between the *Elements* and what went before poses a problem. According to Proclus, Euclid 'brought

[1] *Commentary on the first book of Euclid's Elements*, p 68 17 ff.

together the elements, collecting many of Eudoxus' theorems, perfecting many of those of Theaetetus and providing with incontrovertible demonstration propositions that had been proved less rigorously by his predecessors'. Several writers, beginning with Hippocrates of Chios in the late fifth century, are reported to have composed books of *Elements*, and Proclus also tells us that Archytas of Tarentum and Theaetetus of Athens (both contemporaries of Plato) 'increased the number of theorems and progressed towards a more scientific arrangement of them'. Aristotle explains how the term 'elements' itself was used: 'we give the name "elements" to those geometrical propositions, the proofs of which are implied in the proofs of all or most of the others' (*Metaphysics* 998 a 25 ff). The elements are certain primary propositions from which other propositions may be derived, and we may presume that the general aim of earlier writers on the elements was similar to that of Euclid himself, namely to set out, systematically, a series of fundamental mathematical demonstrations.

Euclid's debts to his predecessors can in many cases be identified. The definition of line as 'breadthless length' (def. 2 of book I) is the same definition that is cited, and criticized, by Aristotle in the *Topics* (143 b 11 ff). A scholiast on book X 9 states that the theorem proved in it was the discovery of Theaetetus and another commentator on that book says that Theaetetus was also responsible for the distinction between three main sorts of irrational lines that we find in it. A scholiast on book V tells us that the general theory of proportion in that book—usually considered one of the finest things in the whole of the *Elements*—was attributed to Eudoxus of Cnidus. An interest in proportion goes back to the very beginning of Greek mathematics. But the great merit of Eudoxus' theory is that it applies to incommensurable as well as to commensurable magnitudes (that is to $1 : \sqrt{2}$, as well as to $1 : 2$) and to magnitudes of any kind (numbers, lines, areas, volumes, times, etc). Again proofs of two of the theorems in book XII are ascribed to Eudoxus by no less an authority than Archimedes. These are XII 7 where it is shown that any pyramid is a third part of the prism that has the same base with it and the same height, and XII 10 which proves that any cone is a third part of the cylinder which has the same base with it and the same height.[1] Elsewhere the debts of other parts of the *Elements* to earlier Greek mathematics—for instance, those of the so-called arithmetical books VII–IX to Pythagorean number-theory—are more difficult to document but hardly less certain.

The obvious question that these observations provoke is what, if anything, is distinctively Euclidean in the *Elements* as we have them? Not many of the theorems and demonstrations that his book contains seem to have been his own discoveries. His own chief contribution consists, rather, in the way the whole work is put together. The *Elements* is a highly systematic work. Book I sets out certain fundamental assumptions and deals with some simple problems in plane geometry (I 47 contains the famous proof of Pythagoras' theorem that the square on the hypotenuse of a right-angled triangle is equal to the sum of the squares on the other two sides), and this discussion continues throughout the next three books, with the help of further definitions as the problems tackled relate to more complex geometrical figures. Book V introduces the theory of proportions which is then applied, in VI, to problems in plane geometry. VII–IX add further definitions concerning numbers and then discuss the nature and properties of whole numbers. X deals with irrationals, and in XI–XIII Euclid turns to problems in solid geometry, using (in the last two books) the method of exhaustion based on the first proposition of book X. In every case except V and VII—which make fresh starts—the later books presuppose and build directly on the conclusions of the earlier. To be sure, certain irregularities and anomalies remain. Thus three of the definitions in book I—those of 'oblong', 'rhombus' and 'rhomboid'—are thereafter nowhere used in the *Elements*. Nevertheless the whole is a highly methodical and coherent presentation of a body of mathematical demonstrations.

More than any earlier work, the *Elements* exemplifies and embodies the notion of an axiomatic, deductive system, but for the origin of that idea itself we must turn back to earlier texts, both mathematical and philosophical. In a famous but obscure passage in the *Republic* (510 b–d) Plato claimed that

[1] According to Archimedes, the propositions themselves were first stated by Democritus, but first rigorously demonstrated by Eudoxus.

the source of all truth and reality is a single first principle—the Form of the Good—which he described as 'unhypothetical' in order to contrast its status with that of the hypotheses or assumptions of mathematics that are provisionally accepted as true but that require confirmation. Aristotle argued, much more clearly, that not all true propositions can be demonstrated. He insisted that the starting-points of demonstrations are principles that are themselves indemonstrable but known to be true, and he distinguished three sorts of such principles, definitions, hypotheses and axioms. Euclid's *Elements* are in fact a series of demonstrations from first principles that are themselves not proved but simply asserted, and he too starts with three types of first principles, similar to, though not identical with, those of Aristotle, namely definitions, postulates and 'common opinions'. The 'common opinions' correspond roughly to Aristotle's axioms, and indeed one of the mathematical examples of an axiom that Aristotle gives reappears in Euclid as the third of his common opinions: 'if equals be subtracted from equals, the remainders are equal.' Euclid's postulates, however, differ from Aristotle's hypotheses, at least as that term is strictly defined in the *Posterior Analytics* 72 a 18 ff. There a hypothesis is an assumption concerning the existence of certain things, the usual example being points and lines; but the first three of Euclid's five postulates are assumptions concerning the possibility of carrying out certain geometrical constructions (for example, 'to draw a straight line from any point to any point') and the last two assume certain truths concerning such constructions, namely that all right angles are equal, and that non-parallel straight lines meet at a point. Whereas the common opinions are self-evident principles that apply to the whole of mathematics, the postulates are the fundamental *geometrical* assumptions underlying Euclid's geometry.

Euclid's general conceptions concerning the form and foundations of an axiomatic system have obvious affinities with those that Aristotle formulated in the context of his study of reasoning in general, although we cannot be sure how far these similarities are due to the direct influence of the one on the other, or how far Euclid is following and developing ideas that were already current among earlier mathematicians.

The particular definitions, postulates and common opinions he proposed throw light on the nature of his mathematics and of Greek mathematics as a whole. Thus his definitions of the unit and of number in book VII show that one was not treated as a number. A unit is that by virtue of which each of the things that exist is said to be one (Def. 1) and a number is a set composed of units (Def. 2). The difference between Euclid and some post-Euclidean mathematics here is not merely one of convention. In Euclid, the one is by implication itself indivisible, and in the arithmetic of book VII fractions are dealt with as ratios or proportions between numbers. To understand the background to this view, we may recall the philosophical problems concerning the one and the many raised by Parmenides and Zeno of Elea in the fifth century. Euclid may have been influenced by arguments of the type reported in Plato (*Republic* 525 de) when he says that certain mathematicians refused to allow the one to be divided, 'lest it should appear to be not one, but many parts'. They argued, it seems, that if the one is allowed to be divisible, the one becomes at the same time many: to avoid the apparent contradiction, the one—like Parmenides' Being—must be indivisible.

The arguments of the Eleatics may again be relevant to our understanding of the fifth common opinion of Euclid. This states that 'the whole is greater than the part'—which seems innocuous enough. We know from Aristotle, however, that one of Zeno's arguments against plurality aimed to prove that 'half of the time equals its double'.

The background of controversy to the famous fifth postulate about parallel lines is more complex. A passage in Aristotle (*Prior Analytics* 65 a 4 ff) shows that current mathematical theory on the subject of parallels was thought open to the charge of circularity, for he remarks that mathematicians who 'think they can construct parallels unconsciously assume such things as cannot be demonstrated if parallels do not exist'. Euclid's position is quite different, in that, having defined 'parallel' in Definition 23 of book I, he adopts as a *postulate* the proposition that non-parallel straight lines meet at a point. There is no evidence that Euclid or any other Greek geometer envisaged the development of other geometries such as those devised since Lobachewsky (born 1792). But it is worth remarking that Euclid's *Elements* are not merely an axiomatic, but also an explicitly hypothetical, system—a system based on postulates and common opinions among which were propositions that he must have known to have been questioned or denied by other Greek philosophers or mathematicians.

Having first set out his assumptions in the form of definitions, postulates and common opinions, Euclid proceeds to the demonstration of his theorems and to the resolving of his problems of construction. Among the methods of argument he uses two are particularly prominent, the so-called method of exhaustion and the more general method of 'reduction to the impossible'. The first was probably the discovery of Eudoxus. The principle that underlies the method is stated at X 1: if there are two unequal magnitudes (A and B) and from the greater (A) there be subtracted more than its half, from the remainder more than its half, and this process is repeated continually, there will be left some magnitude which will be less than the lesser of the two given magnitudes (B). By repeating the process of subtraction as many times as one likes, one will eventually arrive at a remainder that is smaller than any given magnitude at all. One obvious application of this method, in geometry, is to determine a curvilinear area such as a circle. This can be done by inscribing in it regular figures which successively approximate, in area, to the area of the circle itself. Starting with a square (ABCD in Fig. 1) and doubling the sides of the figure on each occasion (i.e. bisecting the arcs AB, BC etc, then the arcs AE, EB etc, and so on), one can increase the area of the inscribed figure until the difference between it and the area of the circle is less than any given magnitude: the area of the circle can then be treated as equivalent to that of the inscribed polygon.

'Reduction to the impossible' or 'proof by the impossible' is named and discussed by Aristotle in his logic and this method too had evidently been much used in mathematics before Euclid. In this, the contradictory of the thesis to be proved is assumed and this is shown to lead to impossible or absurd results. It is used, for example, in the proof of the proposition that the number of primes is infinite (IX 20). This proceeds by first assuming that the number of primes is finite and then demonstrating the falsity of this assumption: he shows that if we assume that the prime numbers are a finite set, A, B, C, ... X, then the number formed by the product of these numbers plus one $(A \times B \times C \ldots \times X) + 1$ either is itself, or is divisible by, a new prime number not in the set. It was for such simple and rigorous demonstrations that the *Elements* deservedly became famous and achieved such an astonishing success as a text-book: works based on translations of Euclid continued to be used to teach mathematics in English schools right down to the present century.

The next thinker whom we must consider, Archimedes of Syracuse, is a far more original mathematician than Euclid

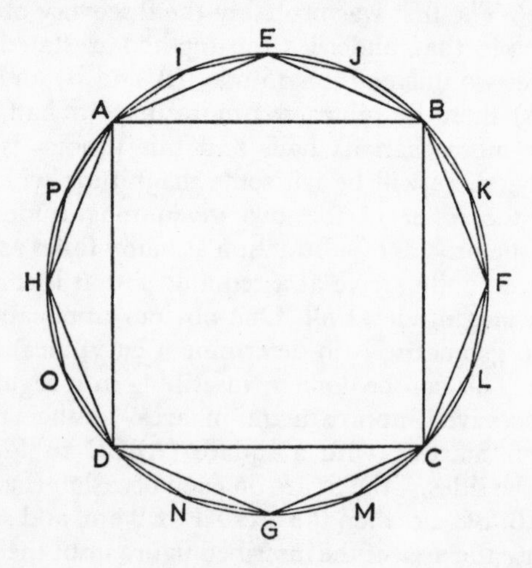

Fig. 1 An application of the method of exhaustion.

and indeed one of the greatest creative geniuses that Greek science produced. We know that he died when Syracuse was taken and sacked by Roman troops under Marcellus in 212 B.C., and if we accept an admittedly late report that he was 75 years old at the time, he was born in 287. In the *Sand-Reckoner* he includes his own father, Pheidias, among the 'earlier astronomers'—along with Eudoxus and Aristarchus—and as I have already noted (p 6) our sources tell us that Archimedes was the friend and kinsman of the ruling family at Syracuse. He is reported to have visited Alexandria, and he certainly knew and corresponded with Eratosthenes. But most of his life seems to have been spent in his home town of Syracuse.

The range of his interests was wide, comprising not only arithmetic and geometry, but also optics, statics and hydrostatics, astronomy and engineering. Several of the stories told about him in our secondary sources relate to his skill as an engineer, and I shall be discussing this in Chapter 7. But apart from doubtful references in Arabic sources to a book on water-clocks, the only written work that he composed on a technological subject was the treatise (not extant) *On Sphere-Making* which dealt with the problem of constructing a sphere to represent the movements of the sun, moon and planets—Archimedes himself evidently made such a planetarium which survived to the time of Cicero who refers to it on several occasions in terms that convey his admiration for Archimedes' skill but that do not, unfortunately, provide much information about the nature of the sphere or how it worked. An optical treatise and several books on arithmetic, geometry and statics (for example *On Balances*) have also been lost. Even so we have no less than nine complete or nearly complete works as well as fragments, or Latin or Arabic translations, of several others.

As an example of his work in arithmetic, we may take the treatise entitled *The Sand-Reckoner* which discusses certain problems involved in dealing with very large numbers. In this Archimedes sets himself the task of calculating the number of grains of sand that the universe (that is, the sphere of the fixed stars) would hold on three principal assumptions: (i) that a sphere of diameter $\frac{1}{40}$th of a finger's breadth would hold no more than 10,000 grains of sand; (ii) that the perimeter of the earth is not more than 3,000,000 stades; and (iii) that the diameter of the sphere of the fixed stars bears the same relation to the diameter of the sphere whose radius is the line joining the sun and the earth as that diameter does to the diameter of the earth. Further assumptions and calculations yield a fourth proposition that the circumference of the sun's orbit round the earth is less than 30,000 earth diameters. In the introduction to the work he refers to Aristarchus' heliocentric theory (on which see below, pp 53 ff). Archimedes mentions this view not because he endorses it (in fact he does not), but because Aristarchus had suggested that the diameter of the earth's orbit round the sun bears the same proportion to the distance of the fixed stars as the centre of the sphere bears to its surface. But since the centre of a sphere has no magnitude, Aristarchus' statement implies that the sphere of the fixed stars is infinite—which would make Archimedes' problem of *numbering* the grains of sand impossible. For the sake of his problem, then, Archimedes adopts the assumption in the form set out as (iii) above. The figure he arrives at for the diameter of the sphere whose radius is the line joining the earth and the sun is not greater than 10^{10} stades, and the figure for the diameter of the sphere of the fixed stars is not greater than 10^{14} stades.

Although Archimedes describes certain astronomical observations that he carried out to check the angle that the diameter of the sun subtends at the eye, it is clear that in this treatise he is interested in the astronomical data only in so far as he has to use them to specify the conditions of his problem. He was fully aware that the perimeter of the earth is nowhere near as great as 3,000,000 stades and he mentions the fact that some astronomers had calculated it at about 300,000 stades. He assumes a much higher figure in order to make the conditions of the problem as stiff as possible. When he proceeds to its solution, he first sets out the notation he has invented to describe very large numbers. Greek mathematical notation has often been criticized for its clumsiness—and with some reason, although neither the poverty of symbols for multiply, divide, equals, proportional to and so on, nor the alphabetic system of numeration (in which α stands for 1, β for 2, γ for 3, ι for 10, $\iota\alpha$ for 11, κ for 20, ρ for 100 and so on) prevented the carrying out of complex operations such as multiplication and division involving extremely large numbers or complicated fractions. Yet we find Archimedes devising a notation that allows him to name numbers up to the number we should represent by 1 followed by 80,000 million million ciphers—that is $10^{8 \cdot 10^{16}}$—a number which he describes, with striking economy, in seven Greek words, *hai mȳriakismȳriostas periodou mȳriakismȳriostōn arithmōn mȳriai mȳriades*, literally a myriad myriad units of the myriad-myriadth order of the myriad-

myriadth period. Archimedes is then able to show that—given his original assumptions—the number of grains of sand that the sphere of the fixed stars would hold can easily be expressed in this notation. The actual solution he arrives at is that the number in question is not greater than the number we should express as 10^{63}.

Archimedes' primary interest was in geometry, to which most of the extant treatises are devoted. In the short work *On the Measurement of the Circle*, for instance, he determines the area and circumference of a circle and gives an arithmetical approximation to the value of π, namely $3\frac{1}{7} > \pi > 3\frac{10}{71}$. Elsewhere he tackles such problems as determining the area of

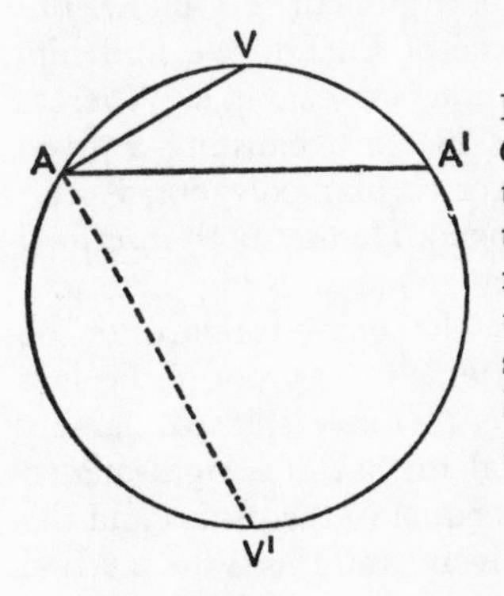

Fig. 2 Archimedes, *On the Sphere and Cylinder* I Proposition 42 proves that the surface of the segment AVA′ of the sphere is equal to the circle whose radius is VA (V being the vertex of the segment). I Proposition 43 proves similarly for segments greater than a hemisphere that the surface of the segment AV′ A′ is equal to the circle whose radius is V′A.

a parabolic segment or that of a spiral, or finding the surfaces and volumes of a sphere or a segment of a sphere. In the treatise *On the Sphere and Cylinder*, for example, he demonstrates the following theorems among others: (1) that the surface of a sphere is four times that of a great circle of the sphere (that is, surface $= 4\pi r^2$); (2) that the surface of any segment of a sphere is equal to that of a circle whose radius is equal to the straight line drawn from the vertex of the segment to any point on the circumference of the circle which is the base of the segment of the sphere (see Fig. 2); (3) that the volume of a cylinder whose base is equal to the great circle of a sphere and whose height is equal to the diameter of the sphere is half as large again as the volume of the sphere (this gives the volume of a sphere as $\frac{4}{3}\pi r^3$); and (4) that the surface of the circumscribing cylinder including its bases is half as large again as the surface of the sphere. Archimedes is careful to distinguish his own work from that of earlier mathematicians and to acknowledge his debts to them. But he states in the introduction to this treatise that these are all new theorems, which had never before been demonstrated.

The style of presentation in the geometrical treatises is similar to that of Euclid's *Elements*. He begins, where necessary, with a statement of definitions and postulates, proceeding then to the proof of his theorems in orderly sequence, although, unlike Euclid and thanks largely to his work, Archimedes is able to take the proof of many elementary theorems in geometry for granted. His methods of argument too follow, but in some respects advance on, those of Euclid. We find him modifying the method of exhaustion, for example. Whereas Euclid had tended to confine his use to exhausting a given area by *inscribing* successively larger regular polygons, Archimedes' attack is often double-pronged. He uses both inscribed and *circumscribed* figures with a view to compressing them to a point where they coalesce with the curved figure to be measured. This is the method on which the proof of the first proposition in *On the Measurement of the Circle* is based, namely that the area of the circle is equal to that of a right-angled triangle with the two smaller sides equal to the radius and the circumference of the circle respectively, and this is the method that gives the upper and lower limits within which π must fall.

A more strikingly original aspect of Archimedes 'methods is his application of mechanical concepts to geometrical problems. Thus arguments based on principles derived from statics (in particular, the law of the lever) are brought to bear on the problems of determining unknown areas or volumes. By thinking of a plane figure as *composed of a set of parallel lines indefinitely close together*, and then thinking of these lines as *balanced* by corresponding lines of the same magnitude in a figure of known area, one can find the desired area in terms of the known area. Similarly by thinking of a solid as made up of a set of planes and by thinking of these planes as balanced by corresponding planes in a solid of known volume, one can find an unknown volume in terms of a known one.

In a passage in the incomplete work entitled *The Method*,

addressed to his contemporary Eratosthenes, Archimedes refers to this method:

> I thought fit to write out for you and explain in detail in the same book the peculiarity of a certain method, by which it will be possible for you to get a start to enable you to investigate some of the problems in mathematics by means of mechanics. This procedure is, I am persuaded, no less useful even for the proof of the theorems themselves; for certain things first became clear to me by a mechanical method, although they had to be demonstrated by geometry afterwards because their investigation by the said method did not furnish an actual demonstration. But it is of course easier, when we have previously acquired, by the method, some knowledge of the questions, to supply the proof than it is to find it without any previous knowledge.[1]

He gives as an example what he describes as 'the very first theorem which became known to me by means of mechanics', namely the theorem that any segment of a parabola is four-thirds of the triangle that has the same base and equal height. But then having set out how he discovered this theorem by the mechanical method, he goes on: 'now the fact here stated is not actually demonstrated by the argument used; but that argument has given a sort of indication that the conclusion is true.' And he then refers to the proof that he had given of this theorem in *On the Quadrature of the Parabola*, where indeed he offers two demonstrations, one using mechanical methods, the other purely geometrical ones.

As Heath, for example, has pointed out, it is remarkable that whereas a modern mathematician would consider the mechanical argument adequate proof of the theorem in question, Archimedes insists that his mechanical method is not a method of proof, only one of discovery. Indeed it was only because he believed that the method might prove useful to other researchers that we are told how it was that he had found the theorem in the first place. Normally all trace of the process of discovery has been removed from Greek mathematical texts. In general, all that Greek mathematicians were

[1] From the translation in T. L. Heath, *The Works of Archimedes* (Cambridge University Press, 1912; Dover Books (no date)).

concerned to publish—that is to communicate to their colleagues—was the final product of their research, in the form of a group of theorems rigorously demonstrated and set out as a deductive system. What we are given is what is known as the 'synthesis', and no trace of the preceding 'analysis'[1] or of any process of discovery is preserved. Archimedes' account of his mechanical method is an exception to the general rule, but it is clear that he too shares the usual view that the discovery of the theorem is of minor importance compared with that of its proof.

The use of mechanical concepts is a distinctive feature of Archimedes' geometry. Conversely, his treatment of certain problems in mechanics is geometrical. The best examples of this come from his statics and hydrostatics, fields in which his work was, once again, highly original. Although there are earlier discussions of the law of the lever in, for example, *On Mechanics*, a treatise by a follower of Aristotle found in the Aristotelian corpus, Archimedes was the first to demonstrate and systematize the elementary theorems of statics in his book *On the Equilibrium of Planes*, while in hydrostatics he had, so far as we know, no predecessors of any note at all.

The contrast between the story in Vitruvius of how Archimedes solved the problem of the crown and the extant hydrostatical books themselves is instructive. In his *On Architecture* (IX, Preface 9 ff), Vitruvius relates that Archimedes was asked by Hiero to discover whether or not a crown that

[1] The nature of Greek geometrical analysis is disputed, though it is clear that it begins by assuming what is required and then seeks its proof. On one view it does so by proceeding deductively, considering the consequences of what has been assumed. But on another it consists rather in finding the premisses from which the given conclusion follows. Thus Pappus (*Mathematical Collection*, VII, 1, 634 13 ff, trans. T. L. Heath, *A History of Greek Mathematics*, Oxford, Clarendon Press, 1921, Vol 2) includes this passage: 'we assume that which is sought as if it were already done, and we inquire what it is from which this results, and again what is the antecedent cause of the latter, and so on, until by so retracing our steps we come upon something already known or belonging to the class of first principles.' Synthesis then proceeds in the reverse direction, taking the antecedents and consequences 'in the natural order'—that is in the order in which they are presented in the proof.

had been made for him was pure gold. The weight of the crown corresponded to the weight of gold that the contractor had received, but had the contractor adulterated the gold with silver? Archimedes is said to have discovered how to solve the problem in the bath, when he observed that the lower his body sank into the water, the more the water overflowed the bath. This gave him the idea of making two other masses, one of gold, the other of silver, of the same weight as the crown, and of measuring the amount of fluid each displaced when immersed in a vessel full of water: if the crown displaced more water than the equivalent weight of gold, this would indicate the presence of an alloy. Whereupon as every schoolboy knows, Archimedes jumped out of the bath and ran home naked shouting '*Eureka*—I have found it'.

This story is open to question at several points. Even the report of the method used to demonstrate the proportion of the alloy may be incorrect. Rather than measure the amount of water displaced by the three masses, Archimedes may simply have weighed them in water and noted the apparent loss of weight in each case. This too would have revealed the difference in the specific gravities of the three masses and the proportion of the alloy, and indeed, unlike the first way, it does so by a method that incorporates the principle named after Archimedes. Nevertheless, even if Vitruvius' story is a fabrication, it captures the excitement of the moment of discovery of the solution to a problem. But then the contrast between the story and Archimedes' own treatise *On Floating Bodies* is remarkable. Here the excitement of discovery is replaced by the cool logic of geometrical demonstration. The work begins, in true Euclidean fashion, with the first postulate, a statement concerning the properties of fluids that have to be assumed. Book I demonstrates, for example, that 'any solid lighter than a fluid will, if placed in the fluid, be so far immersed that the weight of the solid will be equal to the weight of the fluid displaced' (Proposition 5, Heath's paraphrase), and the proposition that contains the so-called principle of Archimedes itself:

> solids heavier than the fluid will, if placed in the fluid, be carried down to the bottom of the fluid, and they will be

lighter in the fluid by the weight of the amount of fluid that has the same volume as the solid. (Proposition 7.)

The proof of this last proposition is simple enough, but thereafter the propositions demonstrated, and the geometry used in the demonstrations, become increasingly complex. After dealing with two further propositions concerning segments of a sphere in book I, Archimedes turns, in book II, to problems concerning paraboloids of revolution (the figure generated by a parabola rotated on its axis), investigating the conditions of stability of segments of paraboloids of varying shapes, and of varying specific gravities, in a fluid.

The treatise on statics, *On the Equilibrium of Planes*, takes a similar form. The use of the balance and of levers of various types had long made men familiar with certain fundamental properties of the lever. *On Mechanics* had discussed such practical problems as why larger balances are more accurate than smaller ones and why longer bars are moved more easily than shorter ones round the same capstan. What is new in Archimedes' work is the rigorous deductive proof of the basic propositions of statics. He begins by stating seven postulates, for example that 'equal weights at equal distances are in equilibrium, and equal weights at unequal distances are not in equilibrium but incline towards the weight which is at the greater distance' (Postulate 1). A series of simple propositions are then proved, mostly by 'reduction to the impossible'. This prepares the way for the proof of the law of the lever in the two fundamental propositions (6 and 7) where he establishes first for commensurable, and then for incommensurable, magnitudes, that two magnitudes balance at distances that are reciprocally proportional to the magnitudes. The rest of book I and book II deal with the problems of determining the centres of gravity of various plane figures, such as a parallelogram, a triangle and a parabolic segment. In complete contrast to the repeated references to empirical data in *On Mechanics*, Archimedes' work deals with statical problems formulated in ideal, mathematical terms. Friction, the weight of the balance itself, indeed every extraneous physical factor is discounted. The treatment is geometrical throughout, and the whole is an exercise in deductive proof on the model of Euclid's

Elements and itself a model, in turn, of the application of mathematics to physical problems.

Two other important Hellenistic mathematicians may be discussed more briefly, namely Eratosthenes of Cyrene and Apollonius of Perga. Eratosthenes, who was, as we have seen, a contemporary and acquaintance of Archimedes, won a reputation in a large number of fields, mathematics, astronomy, geography, music, philosophy, philology and literature. For this he was nicknamed Pentathlos and Beta—to insinuate that he was not supreme in any field. Ptolemy Euergetes nevertheless invited him to teach his son Philopator and made him head of the Library at Alexandria. Unfortunately we depend entirely on secondary sources for information about him. So far as his scientific work goes, we know that he suggested a new solution to a problem that had exercised mathematicians since the fifth century, that of duplicating a cube, and he is also credited with what is admittedly a crude method of finding prime numbers, the so-called sieve named after him.

But his most interesting work, from our point of view, is in the application of mathematics to geography. Greek geographers had long divided the world into zones, but Eratosthenes' was the first detailed map of the world based on a system of meridians of longitude and parallels of latitude. He was also responsible for a close approximation to the dimensions of the earth, based on observations made at Syene (near the modern Aswan) and Alexandria. At Syene, an upright gnomon cast no shadow at noon on the day of the summer solstice, whereas at Alexandria at the same moment a gnomon made a shadow of $7\frac{1}{5}°$. By simple geometry, this gives a similar figure for the angle subtended by the arc Syene–Alexandria at the centre of the earth (see Fig. 3). Eratosthenes assumed that Syene and Alexandria are on the same meridian (though in fact Syene is 3° east of Alexandria) and he took the distance between the two points to be 5000 stades. This gives a figure of 250,000 stades for the circumference of the earth, although other sources report that he actually adopted a figure of 252,000 stades—whether as a result of fresh observations or, more probably, simply in order to have a more convenient figure for subdivisions of the circumference. Stades of widely differing values are known from antiquity and we cannot be certain which Eratosthenes was using. It has commonly been supposed that he used a stade of 157·5 metres which gives a figure for the polar circumference of 39,690 kilometres, as compared with the modern figure of 40,009 kilometres, but there are at least two other distinct possibilities—stades of $\frac{1}{8}$th and $\frac{1}{10}$th of a Roman mile, that is approximately 186 and 148·8 metres—which give results a good deal less close to the true figure, about 17% too high or 6% too low. But the

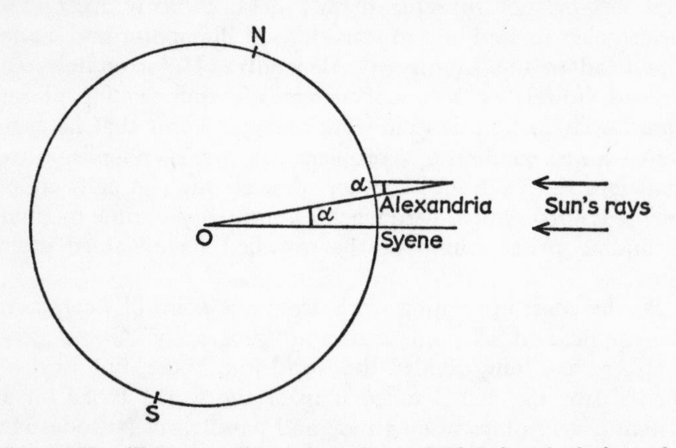

Fig. 3 To illustrate Eratosthenes' method of calculating the circumference of the earth. The two angles marked α are equal.

accuracy of the result he obtained is less important than the method he used to obtain it.

Apollonius, finally, was slightly younger than Eratosthenes and Archimedes, the period of his main activity falling between about 220 and 190 B.C. He was born at Perga in Pamphylia and visited Ephesus and Pergamum: we also hear of him at Alexandria, where indeed much of his work was probably done. Several mathematical and astronomical treatises are mentioned in our sources, but the one major work of his to have survived is his treatise *On Conics*. This was in eight books, of which the first four have been preserved in Greek, and V–VII in an Arabic translation—book VIII is lost.

This is one of the masterpieces of Greek mathematics. The

subject of conic sections had been discussed by several earlier mathematicians: we hear of a treatise by Euclid, for example, on the topic. But Apollonius' work was the first comprehensive and systematic treatment of it. The very terms by which the three types of conic sections are known, ellipse, parabola, and hyperbola, derive from him. Their pre-Apollonian names correspond to the way in which the curves were originally obtained: they were thought of as sections of three kinds of right circular cone, namely cones in which the vertical angle is acute, right-angled and obtuse respectively, the section, in each case, being a plane perpendicular to a generator of the cone (see Fig. 4). The new Greek terms, ellipse, parabola and hyperbola, coined by Apollonius, are introduced in the first

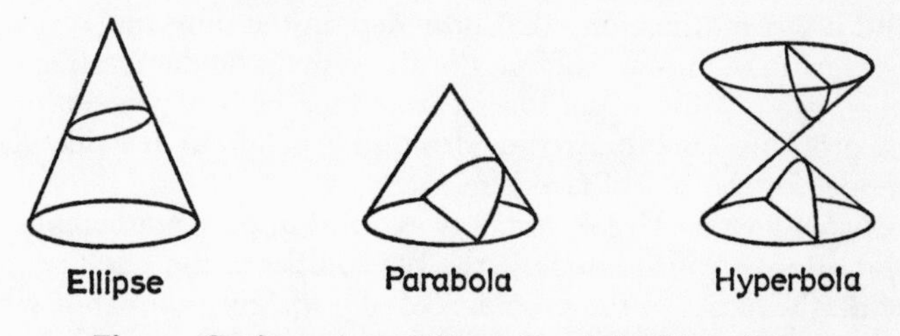

Fig. 4 Conic sections: Ellipse, Parabola, Hyperbola.

few propositions of book I and derive from the constructions which indicate the fundamental properties of the three types of curve. After demonstrating these, Apollonius proceeds to a full investigation of problems connected with conics, the construction of conics from certain data, tangents, focal properties, intersecting conics and so on.

The work of mathematicians of the third and second centuries B.C. is perhaps the greatest—certainly it was the most permanent—achievement of Greek science. In earlier centuries the role of mathematics and its relation to physics and philosophy were, to a large extent, uncertain, as we can see, for example, from the suggestive, but obscure, doctrine that 'all things are numbers' that Aristotle attributed to the Pythagoreans.[1] But after Plato and more particularly after

[1] See *Early Greek Science*, pp 25 ff.

Aristotle himself mathematics achieved a greater independence from philosophy and reached maturity in a series of brilliant works. To be sure, many of the underlying assumptions of Greek mathematics continue, as we have seen, to reflect philosophical controversies, and while some of its methods of argument are strictly mathematical, others, naturally, are common to deductive logic in general. But mathematics was more clearly defined as a separate discipline, and this is reflected in the greater degree of specialization of those who engaged in it in the third and second centuries. Their interests were more often confined to the various branches of mathematics itself—to the exclusion of such subjects as the constitution of matter or the classification of natural substances. Plato had suggested that mathematics is subordinate to dialectic. But it was mathematics that provided, in the third and second centuries, the finest examples of the systematic demonstration of a body of knowledge. The clear and methodical presentation of proof in Euclid, Archimedes and Apollonius became the model for the rest of Greek science.

Moreover the Greeks not only excelled in pure mathematics, but also recognized some of the possibilities of the application of mathematics to the problems of physics, even though it was only in some comparatively simple areas that they did so. There is a striking contrast between the bold but purely speculative ideas of Epicurus and the Stoic philosophers on such physical questions as the constitution of matter, and the lasting results obtained by Archimedes—notably the law of the lever and the law of floating bodies—by applying mathematics to physical problems. The examples we have discussed come from statics, hydrostatics and geography. Others could be given from optics and harmonics. More important, however, is the application of mathematics to astronomy, and this demands separate consideration.

5
Hellenistic Astronomy

THE dominant model in astronomical theory down to the end of the fourth century B.C. was Eudoxus' doctrine of concentric spheres. In this the complex movements of each of the planets, the sun and the moon were represented as the product of the combination of the simple circular movements of a number of concentric spheres. Eudoxus, Callippus and Aristotle differed in their views on the number and nature of the spheres, but all three adopted some version of this hypothesis.[1] The earth was imagined as at rest in the centre of the system. This was, of course, the usual belief: yet there had been, and there continued to be, exceptions. Some of the fifth century Pythagoreans had made the earth a planet, maintaining—largely for symbolic or religious reasons—that the centre of the universe was occupied by an invisible central fire. And in the fourth century Heraclides of Pontus, while believing that the earth is in the centre, thought that it might be in movement. He suggested that certain phenomena may be explained on the hypothesis that the earth rotates on its axis once every twenty-four hours. In the third century two particularly striking ideas were introduced into astronomy, the heliocentric hypothesis of Aristarchus of Samos, and the twin models of epicycles and eccentric circles of Apollonius of Perga.

Our information concerning Aristarchus' theory is second-hand. The only work of his that has been preserved is the short treatise *On the Sizes and Distances of the Sun and Moon*, and this contains no hint of the heliocentric theory—nor indeed was there any need to mention it, since the investigation of the sizes and distances of the sun and moon is independent of whether the sun or the earth is taken as the centre of the planetary system. But although we have no text of Aristarchus himself on the subject, the evidence that he put forward the heliocentric hypothesis is unimpeachable. As we saw in the

[1] See *Early Greek Science*, ch 7.

preceding chapter, Archimedes, a slightly younger contemporary of Aristarchus, refers to this theory in the introduction to his work *The Sand-Reckoner*:

> Aristarchus of Samos brought out a book of certain hypotheses, in which it follows from what is assumed that the universe is many times greater than that now so called. He hypothesizes that the fixed stars and the sun remain unmoved; that the earth is borne round the sun on the circumference of a circle . . .; and that the sphere of the fixed stars, situated about the same centre as the sun, is so great that the circle in which he hypothesizes that the earth revolves bears such a proportion to the distance of the fixed stars as the centre of the sphere does to its surface.

As we have noted, Archimedes remarks on this last suggestion that 'since the centre of the sphere has no magnitude, no more can we conceive it to bear any ratio whatever to the surface of the sphere'. But it is evident that Aristarchus' original intention at this point was merely to insist that the sphere of the fixed stars is immeasurably greater than the sphere of the earth's orbit round the sun.

The theory as described by Archimedes is clear. First, instead of conceiving the sphere of the fixed stars as making a revolution of the heavens once every twenty-four hours, Aristarchus —like Heraclides—held that the sphere of the fixed stars remains unmoved and that the earth rotates on its axis. Secondly, instead of picturing the sun and the rest of what we call the solar system as moving round the earth, he maintained that the earth, and presumably the planets also, move round the sun. The idea that Venus and Mercury move in circles round the sun may well have been put forward before, although the evidence connecting this doctrine with Heraclides in particular is weak. But whereas the fact that Venus and Mercury always stay close to the sun is a matter of simple observation (they are now morning, now evening, stars), there is no such obvious immediate link between the superior planets, Mars, Jupiter and Saturn, and the sun. Aristarchus' originality lies in his making the sun the centre of the whole system. We do not know how far he attempted to work out in detail the movements of the planets, nor even whether he understood that the heliocentric hypothesis provides the basis of the correct explanation of their stations and retrogradations (see below, p 65). Even so, the simplicity and economy that his theory achieved, when compared with Eudoxus' model of concentric spheres, are remarkable. The axial rotation of the earth enabled movements that Eudoxus had attributed to eight separate spheres to be explained as one. Again, whereas Eudoxus had postulated a separate sphere to account for each of the movements of the sun, the moon and the planets along the ecliptic, the heliocentric hypothesis was able to explain these apparent movements on a unitary assumption.

But whereas the main elements of Aristarchus' theory are not in dispute, there is some doubt about his intention in putting it forward. Archimedes says that he 'brought out a book of certain hypotheses'. But Plutarch, writing more than 300 years later, reports (*Platonic Questions* VIII, 1, 1006 c) that the idea that the earth 'turned and revolved' was maintained by both Aristarchus and Seleucus of Seleucia, 'the former merely hypothesizing this, the latter also asserting it'. The nature of the distinction that Plutarch draws between Aristarchus and Seleucus here, and the evidence on which it is based, are unclear. It may be that Plutarch was simply relying on the word 'hypothesis' used by Archimedes, but our first task, in any event, is to determine the force of the term as applied to Aristarchus by Archimedes.

As we have seen, Greek mathematicians habitually begin their works with a statement of assumptions, 'definitions', 'axioms' or 'common opinions', and 'postulates' or 'hypotheses'. Euclid's *Elements* begin in this way, and so too do several of Archimedes' works. Moreover the sole surviving treatise of Aristarchus also starts by setting out certain hypotheses—for example that 'the moon receives its light from the sun'—and this makes it all the more likely that the book that contained the heliocentric theory took a similar form. If so, the hypotheses mentioned by Archimedes, that 'the fixed stars and the sun remain unmoved', 'that the earth is borne round the sun' and so on, were the premisses on which certain conclusions were based, but were themselves adopted without proof.

Furthermore Greek mathematicians did, on occasions, adopt hypotheses that they knew to be false. Thus we have seen that in the *Sand-Reckoner* Archimedes took a figure for the circumference of the earth that is vastly in excess of the one he knew to be approximately correct in order to make the conditions of his problem as stiff as possible. Two of Aristarchus' own hypotheses in *On the Sizes and Distances of the Sun and Moon* are also assumed for the sake of the calculations in that work. Thus he hypothesized that 'the earth bears the relation of a point and centre to the sphere of the moon'. This simplifies his calculations by ignoring parallax, that is the effects that are due to the fact that the position of the observer is on the surface of the earth, not at its centre. Secondly—and much more strikingly—he assumed that 'the moon subtends one fifteenth part of a sign of the zodiac', that is 2°, a figure far in excess of the true value of about ½°. Reasonable approximations to the angular diameters of the sun and moon had been made long before Aristarchus. Indeed another passage in Archimedes implies that Aristarchus himself assigned a value of ½° to the angular diameter of the moon, and even though the work *On the Sizes and Distances of the Sun and Moon* may, of course, have been written *before* Aristarchus arrived at that figure, it still seems unlikely that at any time in his life he believed that the figure of 2° was exactly, or even roughly, correct. It is more probable that this value was hypothetical in our sense of that term, that is, that it was adopted merely for the sake of argument. We naturally expect that the chief purpose of a treatise with the title and form of Aristarchus' will be to arrive at certain concrete results, namely estimates of the actual sizes and distances of the sun and moon. Yet Aristarchus' own main motive was, almost certainly, rather different. What he was interested in was, rather, the mathematical exercise, that is in solving the geometrical problems that the question presented. He deals with the question of determining the sizes and distances of the sun as, first and foremost, a problem in deductive reasoning, and the accuracy of the figure he took for the diameter of the moon was of less concern to him than the solution of the geometrical problem.[1]

[1] Thus Aristarchus presents his results in the form of proportions or ratios, not in terms of absolute figures, for example 'the distance

What, then, was the status of the heliocentric hypothesis? Clearly it does not follow from its inclusion among the hypotheses of his book that Aristarchus would, if pressed, have affirmed that the sun is the centre of the universe. If Plutarch is to be believed, Aristarchus might not have done so, whereas Seleucus would, though we do not know whether Plutarch had independent grounds for his opinion. But although we must withhold judgement on that point, there is no reason to doubt that the heliocentric hypothesis was put forward, in all seriousness, as one possible mathematical model to account for the movements of the heavenly bodies. This is clear from the way in which the hypotheses are framed and in particular from the care with which Aristarchus guarded against one major objection to which the heliocentric theory was vulnerable, namely the absence of stellar parallax. If the earth moves round the sun, then there should be some variation in the relative positions of the stars observed from different points in the earth's orbit—whereas no such variation was observed in antiquity.[1] But Aristarchus saw that this objection is not valid if the stars are sufficiently far away from the earth. He did not attempt to *prove* the immense distance of the stars. Rather, he included as one of his initial assumptions that 'the sphere of the fixed stars . . . is so great that the circle in which . . . the earth revolves bears such a proportion to the distance of the fixed stars as the centre of the sphere does to its surface'. If the stars are at infinite distance, then no stellar parallax would be observed. What Aristarchus invites his fellow astronomers to consider is not merely the bare suggestion that the sun is at the centre of the universe, but a carefully worked out set of interrelated assumptions.

This takes us to the question of why the heliocentric theory had such an unfavourable reception in antiquity. Seleucus of Seleucia—a Chaldaean or Babylonian astronomer who probably worked around the middle of the second century

of the sun from the earth is greater than eighteen times, but less than twenty times, the distance of the moon from the earth' (Proposition 7), and 'the diameter of the sun bears to the diameter of the earth a ratio greater than 19 to 3, but less than 43 to 6' (Proposition 15).

[1] Stellar parallax was not confirmed by observation until the work of F. W. Bessel and others in the period 1835–40.

B.C.—was, so far as we know, the *only* astronomer in antiquity to have adopted Aristarchus' theory. The two greatest names in third and second century astronomy, Apollonius of Perga and Hipparchus of Nicaea, both retained the geocentric view and this doctrine prevailed throughout the rest of antiquity although Aristarchus' theory continued to be mentioned from time to time by various writers.

The reasons for which Apollonius and Hipparchus themselves rejected the heliocentric theory are not recorded. Yet with the help of Ptolemy and other sources we can reconstruct some of the arguments used in this controversy in the third and second centuries. First, it is clear that some lay writers rejected Aristarchus' theory on religious grounds. We have remarked that as early as the fifth century B.C. there were Greek thinkers who removed the earth from the centre of the universe in part because they believed that the earth was not *noble enough* to occupy this position. Nevertheless for most Greeks the idea that the earth is in the centre of the universe was not only a matter of common sense but also a religious belief—one that reflected their conception of the sacredness of earth herself. One writer is known definitely to have spoken against Aristarchus on this score. This was the Stoic Cleanthes, the least original of the three early leaders of a school which, strong as it was in ethics and cosmology, was generally very weak in the special branches of natural science. Plutarch tells us that Cleanthes 'thought that the Greeks ought to indict Aristarchus of Samos on a charge of impiety for putting in motion the Hearth of the Universe [that is, the earth]' (*On the face on the moon*, ch 6, 923a).

Although we have no reason to suppose that the Greeks took Cleanthes' advice and actually prosecuted Aristarchus, many must have been scandalized by his theory. Yet it must be stressed that so far as Aristarchus' fellow astronomers were concerned, the chief grounds for their rejecting the heliocentric doctrine had, so far as we know, nothing to do with religion, but related to the grave astronomical and physical objections to which they considered the doctrine was open.

These objections were of three main kinds. First, there was the Aristotelian argument from natural movements. It is a matter of observation—so this argument ran—that heavy objects naturally travel towards the centre of the earth. Assuming this law applies to heavy objects wherever they may be, the centre of the earth may be presumed to coincide with the centre of gravity of all the heavy constituents in the universe. Moreover once a heavy object reaches its 'natural' place—the place towards which its natural movement is directed—it comes to rest. Applying this idea too to the earth as a whole, we arrive at the conclusion that the earth not only is at rest in the centre of the universe, but also could not be moved except by some force sufficient to overcome its natural tendency.

Secondly, there were arguments based on observations of objects moving through the air. If the earth rotates on its axis or is subject to any movement whatsoever, then—so it was argued—this should have a sensible effect on the movements of objects through the air. Ancient astronomers certainly appreciated that if the earth rotates on its axis every twenty-four hours, the speed of a given point on the surface of the earth must be very great indeed. How, then, could clouds or missiles travelling through the air ever overcome this motion of the earth? They could never achieve any *easterly* movement, for the earth would always forestall them. It is clear from the passage in Ptolemy (*Almagest* I ch 7) reporting this argument that one possible line of defence had already occurred to the ancients, namely that it is not only the earth, but also the surrounding air, that rotates on its axis—a notion that was facilitated by the common Greek distinction between the lower air (*āēr*) and the upper (*aithēr*). Yet, as we shall see, Ptolemy, for one, remained unimpressed by this defence, arguing that even if the air is carried round with the earth, solid bodies moving through the air should still yield evidence of the earth's rotation.

Thirdly, the chief astronomical, as opposed to physical, argument was the difficulty presented by the apparent absence of stellar parallax—an objection that Aristarchus was evidently aware of when he included among his initial assumptions the hypothesis that the stars are infinitely distant from the earth (see above, p 57). Now the absence of stellar parallax presented something of an embarrassment even for those who held that the earth is at rest in the centre of the universe—for why were there no variations in the relative positions of the stars when these were observed from different points on the surface of the earth? Yet in the case of the geocentric theory this was only a minor difficulty. The assumption that the sphere of the fixed stars is incomparably greater than the *earth* was, of course, much easier to make than the assumption that the advocates of the heliocentric theory had to make, namely that the sphere of the fixed stars is incomparably greater than the sphere bounded by the *earth's orbit round the sun*.

These three objections are of very varying force. The Aristotelian doctrine of natural places and movements, while superficially highly plausible, was itself fraught with difficulties. If all 'heavy' objects move towards the same place, how are the movements of the heavenly bodies to be explained? Aristotle had rejected the idea that their motion is in any way forced or artificial, and had concluded that they must be composed of a fifth element, *aithēr*, which is quite unlike the four elements of which physical objects on and around the earth are made in that it is neither hot nor cold, and neither light nor heavy. But this left many fundamental questions unanswered—for example how the region of *aithēr* is related to the sublunary sphere. And yet if the doctrine of aither was rejected, how could the theory of natural movements apply to the stars which behave neither like 'heavy' nor like 'light' bodies on earth? As for the two other main objections to the heliocentric theory, we have seen that the key points in the argument for the defence had already occurred to thinkers in antiquity—even though with the instruments available to them neither stellar parallax, nor the effect of the rotation of the earth on solid bodies moving through the atmosphere, could be observed.

Nevertheless, although none of the objections against the heliocentric theory was conclusive in itself, their cumulative effect was strong enough to persuade the foremost Greek astronomers to reject that hypothesis. We do not know whether Apollonius and Hipparchus were more influenced by physical, or by astronomical, considerations. Later, however, in Ptolemy himself, the physical arguments become paramount. As we shall see in Chapter 8, he preferred the geocentric view primarily for reasons connected with his Aristotelian physical doctrines, particularly the doctrine of natural places. Yet he adopted those doctrines not merely because of the authority of Aristotle's name, but also because of their apparent plausibility, their agreement with what observation of moving bodies seemed to suggest to be the case.

In rejecting the heliocentric hypothesis, ancient astronomers may also have been influenced by the fact that by itself it did not help to explain one important and obvious astronomical datum that had already proved a difficulty for Eudoxus' theory. This was the inequality of the seasons measured by the solstices and the equinoxes—where, of course, it made no difference whether the sun or the earth was taken as the centre of the system. The doctrine that was adopted in preference to both Eudoxus' theory of concentric spheres and Aristarchus' heliocentric hypothesis was the twin models of epicycles and eccentric circles. This preserved the two key assumptions, (1) of a geocentric system, and (2) of uniform circular motion,[1] and it had the advantage of comprehensiveness over its rivals. A wide range of astronomical phenomena could be explained quite economically by means of either epicycles or eccentric circles. Either the heavenly body was imagined as moving on a circle (the 'epicycle') whose centre itself moves along the circumference of a second circle (the 'deferent') whose centre is the earth (Fig. 5). Or the heavenly body was assumed to move along the circumference of an 'eccentric' circle, that is one whose centre does not coincide with the centre of the earth (Fig. 6).

In attempting to trace the origin and early development of the use of these two models we are, once again, hampered by the lack of original texts. Although we have Apollonius' *Conics*, we have no astronomical treatise of his. The only extant work of Hipparchus of Nicaea (who worked in the middle of the second century B.C.)[2] is the relatively minor *Commentary on*

[1] Plato is reported to have been responsible for formulating the chief problem of theoretical astronomy as being to explain the apparent irregularities in the motions of the planets in terms of combinations of uniform, circular movements, see *Early Greek Science*, pp 84 f.

[2] Apart from his astronomy, Hipparchus did important work in geography (see D. R. Dicks, *The Geographical Fragments of Hipparchus*,

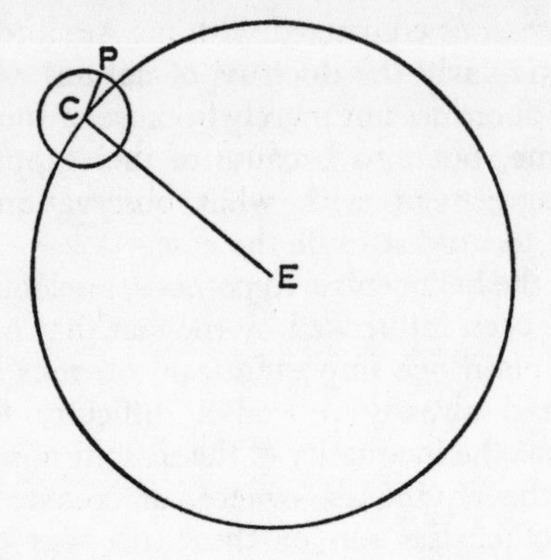

Fig. 5 Epicyclic motion. The planet (P) moves round the circumference of an epicycle, whose centre (C) itself moves round the circumference of the deferent circle, centre E, the earth.

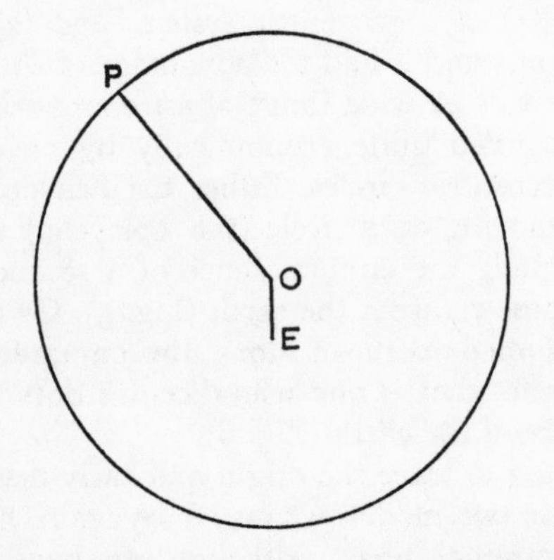

Fig. 6 Eccentric motion. The planet (P) moves round the circumference of a circle, whose centre (O) does not coincide with the earth (E).

University of London, Athlone Press, 1960) and in dynamics: passages from a treatise of his *On Bodies Carried Down by their Weight* are quoted by Simplicius in his *Commentary on Aristotle's On the Heavens* in connection with his discussion, and defence, of Aristotle's views on motion, see further below, pp 157 ff.

the Phaenomena of Eudoxus and Aratus, and even this owes its survival, ironically, to the popularity of Aratus' poem, a third century work based on Eudoxus but itself devoid of any pretensions as an original contribution to astronomy. As so often, Ptolemy proves to be our most valuable source. His own astronomical system incorporates and develops the epicyclic and eccentric models, and he was evidently much indebted to Hipparchus whose work he frequently mentions. He and our other sources cite Apollonius less often, but it is now generally agreed that it is to him and not to Hipparchus that the chief credit for introducing the notion of epicycles and eccentrics into astronomy must go. As we have noted, the idea that Venus and Mercury revolve on circles whose centre is the sun (that is, that they revolve on what came to be known as epicycles) may well have been suggested already in the fourth century. But the first person to attempt to apply the model of epicycles and eccentrics to the problem of the movements of the sun, moon and planets in general was almost certainly Apollonius. Moreover it is likely that he was aware of, and indeed had demonstrated, the geometrical equivalence of the two models, that is he had shown that, if the appropriate parameters are taken, then for every eccentric system one can construct an epicyclic system that will yield exactly equivalent results (Fig. 7 shows the simplest case). Given that this is so, the choice between an eccentric or an epicyclic model will depend, in any particular case, on which of the two provides the simpler solution, that is the one that is mathematically easier to handle.

We can illustrate how the two types of model work in practice with some simple examples. One of the difficulties that Eudoxus' theory encountered was, we said, the inequality of the seasons. Accurate estimates of the lengths of the four seasons were made by Eudoxus' successor Callippus. Starting from the spring equinox, he made them 94, 92, 89 and 90 days respectively, figures which have been calculated to be correct to the nearest whole number for the period at which he was working. But if one assumes that the sun moves round the circumference of a circle whose centre is some distance from the earth, then—still keeping the motion uniform—one can account for the observed data much more simply than on any

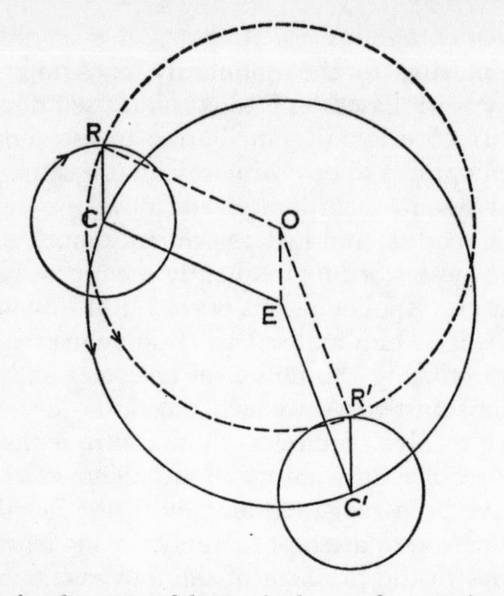

Fig. 7 The simplest case of the equivalence of eccentric and epicyclic motions. When the radius of the deferent circle (CE) is equal to that of the eccentric circle (RO) and the radius of the epicycle (RC) is equal to the eccentricity (OE), then if the angular velocities are regulated so that R and E remain the vertices of a parallelogram CROE, the two models give exactly equivalent results.

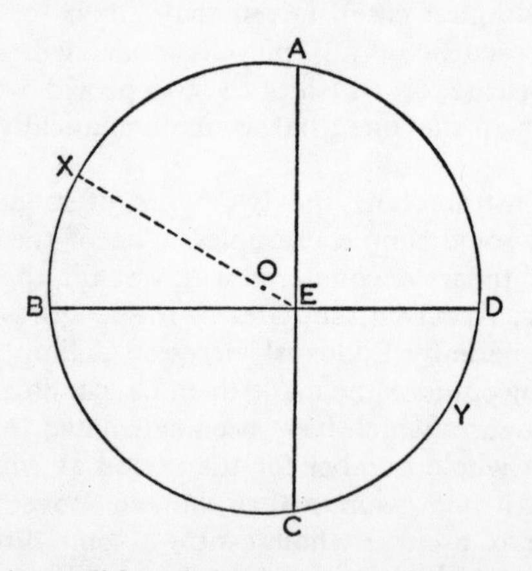

Fig. 8 The inequalities of the seasons explained by the eccentric hypothesis.

theory based on concentric spheres. Fig. 8 shows how this is so. A, B, C, and D are the positions of the sun at the spring equinox, the summer solstice, the autumn equinox, and the winter solstice respectively, E is the earth, O the centre of the circle round which the sun moves and X and Y the positions of the sun at maximum, and at minimum, distance from the earth ('apogee' and 'perigee'). Arc AB (corresponding to spring) > BC (summer) > DA (winter) > CD (autumn). Given estimates of the length of the seasons obtained by observation, the position of the earth (E) can be calculated. Hipparchus' figures have been recorded by Ptolemy (*Almagest* III ch 4). Taking slightly different estimates of the lengths of the seasons from those of Callippus (spring 94½ days, summer 92½) he calculated EO, the distance of the earth from the centre of the circle round which the sun moves, as 'very nearly $\frac{1}{24}$ of the radius of the circle', and he estimated the arc XB at about 24° 30′.

A simple eccentric system provided a neat account of the inequality of the seasons and at the same time explained minor variations in the apparent distance of the sun. But the movements of the moon and the planets are much more complex. The main problem, so far as the planets are concerned was to account for their 'stations' and 'retrogradations'. From time to time the position of a planet in relation to the fixed stars appears to remain the same for a number of days; the planet then appears to move back, from east to west, through the constellations for a time; and then after a second stationary period it resumes its usual easterly movement in relation to the fixed stars. Eudoxus had explained this phenomenon by means of his 'hippopede'—the figure of eight that represents the product of the movement of the two lowest spheres of each planet.[1] But the hypothesis of epicyclic motion can again provide a simpler solution to the problem. In this case the revolution of the planet on the epicycle is in the same sense (not, as with the sun and moon, in the opposite sense) to that of the deferent circle (see Fig. 9 and contrast Fig. 7). When the motions of the epicycle and the deferent act together in the same direction (as at P_1 and P_2 in Fig. 9), the planet will seem

[1] See *Early Greek Science*, pp 86 ff.

to move rapidly through the constellations from west to east. But when the planet comes within the circumference of the deferent, the motion along the epicycle begins to counteract that of the deferent: the planet will first seem to remain stationary, and will then move back through the constellations in a westerly direction (as at P_4 to P_5): there will be a second

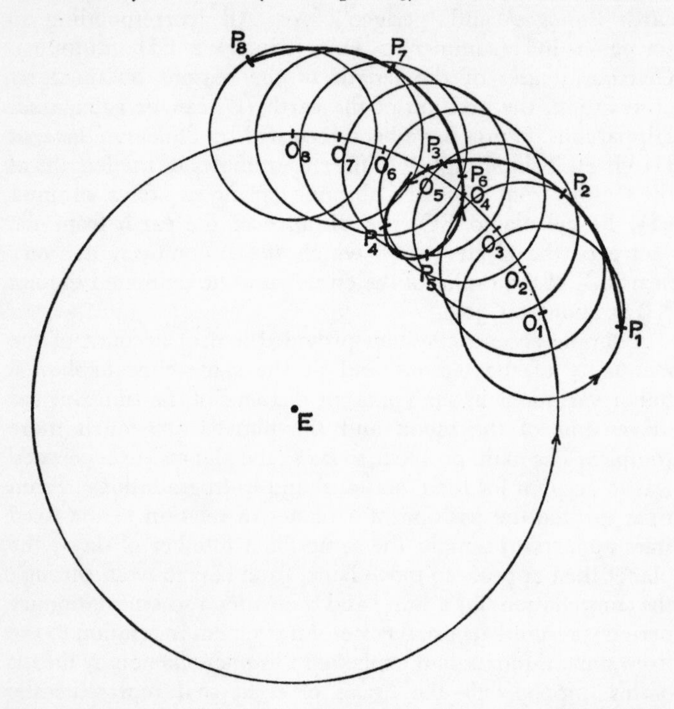

Fig. 9 The epicyclic model used to explain the retrogradation of the planets.

stationary period when the two motions again cancel one another out, and finally, when the two motions act in the same direction again, the planet will resume its usual easterly movement relative to the constellations. Thus the appearances can, in principle, be closely reproduced in a system of great geometrical simplicity.

Between them, the hypotheses of eccentric, and epicyclic,

motion could be adapted to serve as the basis of economical and often tolerably accurate accounts of some highly complex astronomical phenomena. For the motion of the sun, the eccentric hypothesis was preferred. Elsewhere, for the motions of the moon and planets, the epicyclic model was generally used. We do not know how far Apollonius attempted to assign precise values to the eccentric and/or epicyclic motions of the various heavenly bodies, although it is clear from Ptolemy (*Almagest* XII ch 1) that he saw how the epicyclic hypothesis could be used to account for the stations and retrogradations of the planets. But Hipparchus is reported not only to have given a definite estimate of the eccentricity of the sun, but also to have attempted a precise account in the much more difficult case of the moon. So far as the planets are concerned, however, Ptolemy (*Almagest* IX ch 2) tells us that Hipparchus contented himself with assembling more accurate observational data than had been available previously and with demonstrating the inadequacies in current theories of their movements. Nevertheless the flexibility of eccentric and epicyclic models was such that they formed the basis of most subsequent Greek astronomical speculation. In particular, these two types of model, combined with certain other ideas, provided the foundation of the most comprehensive astronomical system elaborated in antiquity, namely that preserved in the *Almagest*, which we shall consider in Chapter 8.

The chief effort of the Hellenistic astronomers was devoted to devising mathematical models to explain the movements of the heavenly bodies. At the same time important work was done in purely observational astronomy. Although the Greeks are still popularly pictured as men who were as negligent in the collecting of data as they were ingenious in constructing theories to explain such data as they had to hand, this is far from the truth, at least so far as Hellenistic astronomy goes.

Hipparchus' achievements in observational astronomy are especially remarkable. Although many aspects of the history of the development of ancient astronomical instruments remain obscure, we can be fairly sure that he was responsible for certain improvements in the basic sighting and measuring instrument, the dioptra. Thus both Ptolemy (*Almagest* V ch 14) and Proclus (*Outline of the Astronomical Hypotheses* ch 4) ascribe to him what they call the 'four-cubit rod' dioptra. The dioptra consisted essentially of a long rod with two sights, one a fixed plate with a pin-point aperture, through which the observer looked, the other a movable plate which was aligned with the target[1] (Fig. 10), although in the more complex version of the instrument described in Hero of Alexandria (*On the Dioptra* ch 3, Fig. 11) the sights are mounted on a bronze disc which

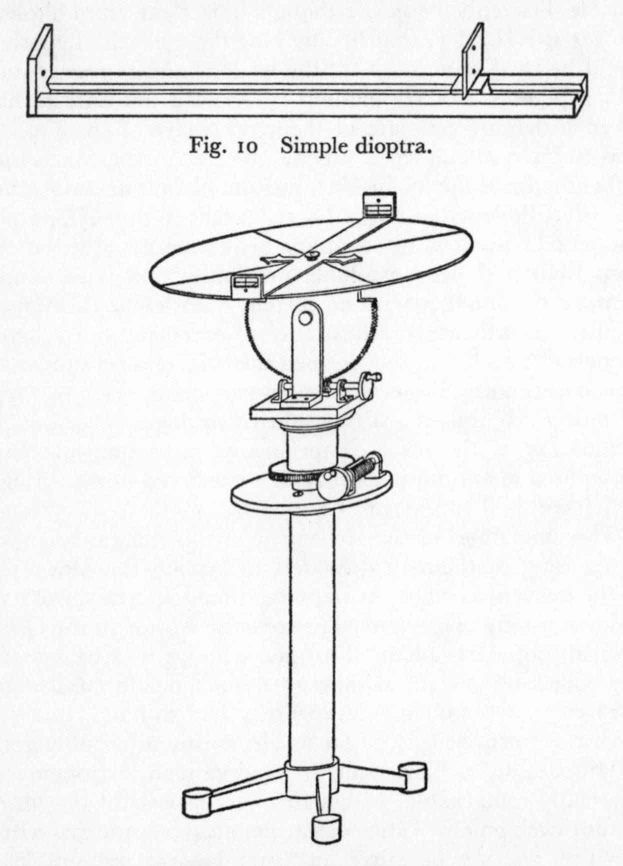

Fig. 10 Simple dioptra.

Fig. 11 Hero's dioptra.

[1] A similar principle is incorporated in the device used by Archimedes to measure the diameter of the sun in the *Sand-Reckoner*.

can be rotated around axes in any direction. It has also been thought likely that Hipparchus already used an armillary astrolabe similar to that described by Ptolemy in *Almagest* V ch 1 (see below, pp 119 f), although conclusive evidence for this is lacking.

What is clear is that Hipparchus undertook a more thorough and detailed survey of the fixed stars than any of his predecessors had attempted. Pliny reports in his *Natural History* (II 24, 95) that

> The same Hipparchus, who can never be praised too highly—for no one has done more than he to demonstrate the kinship of man with the stars and prove that our souls are part of the heavens—detected a new and different star that came into existence in his own time. Its motion where it shone led him to wonder whether this happened at all often and whether the stars which we consider fixed moved. And so he dared to do something that would be rash even for a god, namely to number the stars for his successors and to check off the constellations by name. For this he invented instruments by which to indicate their several positions and magnitudes so that it could easily be discovered not only whether stars perish and are born, but also whether any of them change their positions or are moved and also whether they increase or decrease in magnitude. He left the heavens as a legacy to all mankind, if anyone had been found who could have claimed that inheritance.

Whether or not Pliny is correct concerning the immediate occasion for Hipparchus' undertaking this work—the discovery of a new star—his admiration for the comprehensiveness of Hipparchus' survey is not misplaced. Although the work itself has not survived, it apparently gave the positions of some 850 or more stars in terms of their ecliptic coordinates (longitude and latitude) and it formed the basis of Ptolemy's own star catalogue in the *Almagest* VII–VIII.

Finally Hipparchus' observations led directly to his discovering an important astronomical datum, the so-called 'precession of the equinoxes'. The positions of the equinoctial points (the points where the ecliptic intersects the celestial equator) do not remain constant in relation to the fixed stars, but are displaced from east to west at the rate of about 50 seconds of angle a year. This phenomenon is now explained as being largely due to the fact that the earth is not a perfect sphere, but bulges slightly at the equator. The attraction of the sun and moon tends to pull the equatorial bulge into the

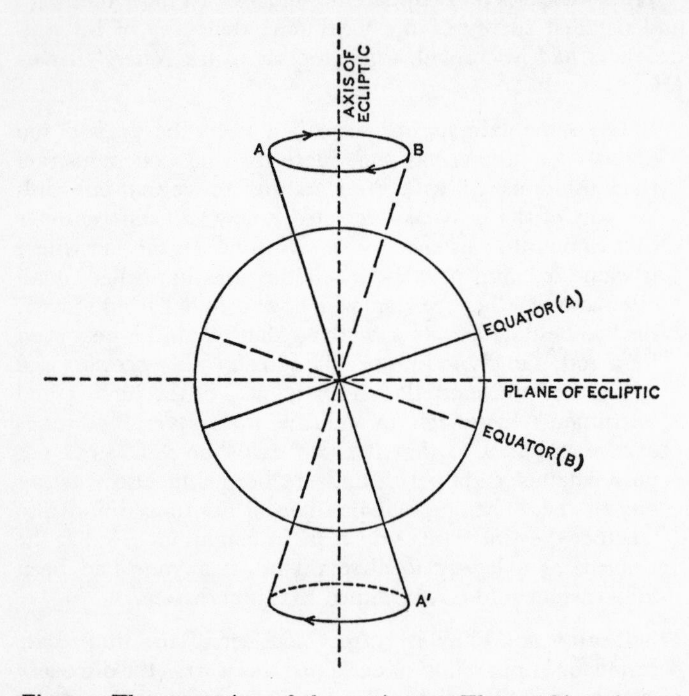

Fig. 12 The precession of the equinoxes. The earth's polar axis (AA′) rotates slowly about the axis of the ecliptic completing a single revolution (A to B back to A) in about 26,000 years.

plane of the ecliptic. This causes the earth's axis to oscillate slightly: it turns very slowly about an axis perpendicular to the earth's orbit, completing a single revolution in a period of about 26,000 years (see Fig. 12).

By comparing his own observations with those made some 160 years earlier by two astronomers called Aristyllus and Timocharis, Hipparchus discovered the change in the relative positions of the equinoxes and the fixed stars, and indeed he gave an astonishingly accurate estimate of the rate of this change. Ptolemy records Hipparchus' discovery in the *Almagest* (VII ch 2):

> For Hipparchus, in his work *On the Displacement of the Solstitial and Equinoctial Points*, comparing the eclipses of the moon, on the basis both of accurate observations made in his time, and of those made still earlier by Timocharis, concludes that the distance of Spica[1] from the autumnal equinoctial point, measured in the inverse order of the signs [that is east to west], was in his own time 6°, but in Timocharis' time 8°, nearly.[2]

Assuming that the movement is uniform, Hipparchus concluded that there was a displacement of not less than 1° in a hundred years, that is 36 seconds of angle a year, a figure that Ptolemy himself was to accept as correct. But if this was the *lower limit* that Hipparchus set for the rate of precession, the data supplied in the passage just quoted suggest that the *actual value* he arrived at may have been even closer to the true figure. Assuming that Timocharis' observations were 160 years before those of Hipparchus, we obtain a figure of 2° in 160 years, or 45 seconds of angle a year, which is within 6 seconds of angle a year of that determined by modern astronomers.

Although the phrase itself is rare in their extant writings, the chief aim of the Hellenistic astronomers was to 'save the appearances' (*sōzein ta phainomena*). This is a complex concept and one that must be carefully distinguished from other contemporary models of explanation such as the Epicurean notion of plural causes. First there is a marked contrast in the matter of determining the 'appearances' themselves. As we have seen, empirical research was foreign to both Epicureans and Stoics and in their astronomical speculations they concentrated very largely on attempting accounts of the most obvious phenomena such as eclipses and the phases of the moon. To be sure, the extent to which the astronomers

[1] A star in the constellation Virgo.

[2] From the translation in T. L. Heath, *Greek Astronomy*, London, Dent, 1932.

themselves carried out systematic observations to verify the phenomena should not be exaggerated. In much of their work the Hellenistic astronomers were content with a very small number of observations from which they extrapolated more general results. A typical instance of this is provided by the evidence in Ptolemy (*Almagest* IV ch 1) that the course of the moon was usually determined from observations of its position during eclipses. Another is that the observations used to discover the lengths of the seasons were probably quite few in number. Moreover we must acknowledge that from the third century onwards much of the stimulus towards the observation of the heavens came from what would nowadays be dismissed as unscientific motives, namely an interest in astrology. Many of the greatest Greek astronomers, including both Hipparchus and Ptolemy, believed that it is possible to predict the future from the stars, in particular by casting horoscopes, and this is probably what Pliny is alluding to when he writes, in the passage just quoted, that no one had done more than Hipparchus to demonstrate the kinship of man with the stars. On the other hand the contrast between the astronomers and the Hellenistic philosophers is sufficiently demonstrated by the evidence of Hipparchus' work in observational astronomy which included not only the discovery of precession, but also, according to the passage in Ptolemy quoted above, p 67, the more accurate determination of the courses of the planets.

But 'saving the appearances' meant not merely suggesting a mathematical model, but suggesting one of a particular sort. The key assumption that underlies the whole of Greek theoretical astronomy is that the irregularities of the movements of the heavenly bodies are to be explained in terms of *regular* and *uniform* motions. The astronomers recognized that certain phenomena, such as the course of the sun, can be accounted for by several alternative mathematical models. But the contrast with Epicurean 'plural causes' is fundamental. The Epicurean doctrine could and often did degenerate into a mere excuse not to subject the proposed account to a critical evaluation (see above, p 25). The astronomers' notion of an explanation was far more stringent, for what they demanded was the simplest possible mathematical solution to which the complex phenomena could be reduced.

Their aims were similar to those of Archimedes, for example, in statics and hydrostatics. Just as, in those subjects, Archimedes treated the problems as problems of applied mathematics rather than of physics, so the astronomers sought a general mathematical theory to explain the movements of the heavenly bodies. As Archimedes ignored the effects of what we should call friction in statics, so the astronomers chose to ignore what they assumed to be minor discrepancies or unimportant details in the data. Thus the orbits of the planets do not coincide exactly with the plane of the ecliptic—yet Greek astronomers often neglected this complication when devising models to explain their motion. Again, although certain physical presuppositions underlie their theories, they made little or no attempt to resolve the *mechanical* problems set by the movements of the heavenly bodies. The stars were often assumed to have their own 'natural' movements, or alternatively (or in addition) they were thought to be alive. But the question of the mechanics of their movements received little attention in the Hellenistic period.

Yet despite the mathematical bias of Greek theoretical astronomy, and its tendency to ignore certain physical aspects of the problems, at a deeper level physical considerations remained important. The unfavourable reception of the heliocentric theory—despite the fact that it was in certain respects obviously far simpler than its rivals—illustrates how the Greeks resisted ideas that conflicted with certain fundamental physical assumptions. Finally, the 'failure' to apply the geometry of ellipses to astronomy also seems surprising but confirms the overriding importance of the principle of uniformity. As we saw in Chapter 4, Apollonius—the very man who was responsible for the model of epicycles and eccentrics—worked extensively on conic sections, including the ellipse. Yet the geometry of ellipses was not applied to astronomy until Kepler. At first sight it might well appear that the Greeks stuck to the assumption of circular motion out of a lack of imagination or sheer stubbornness. But to this it must first be said that the actual differences between the elliptical orbit of the earth, moon and planets and true circular orbits are in some cases only slight (the earth's ellipse has an eccentricity of ·01672). And then a second and more important point is the flexibility of the model of epicycles and eccentrics. By choosing suitable parameters, the epicyclic model can indeed be made to yield any figure, curved or straight, if the speeds with which the epicycle and the deferent revolve are allowed to vary: and even without such variations in speed the epicyclic model can, as Fig. 13 shows, yield an elliptical orbit.

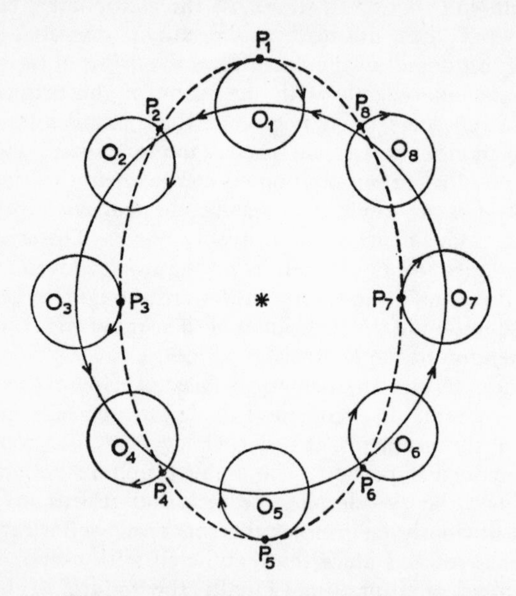

Fig. 13 An ellipse as a special case of epicyclic motion. The planet completes a revolution on the epicycle at the same time as the centre of the epicycle (moving in the opposite sense) completes a revolution round the centre of the deferent.

Given the adaptability of the model they adopted, the adherence to the assumption of circular motion was in line with their general principles of explanation, and in particular with the fundamental principle of uniformity. Circular motion was the simplest and most uniform motion, and it was, as we have seen, axiomatic that the 'appearances' are to be 'saved' by reducing the non-uniform to the uniform.

6

Hellenistic Biology and Medicine

THE lack of reliable first-hand evidence has already obstructed our discussion of Hellenistic mathematics and astronomy. When we turn to biology and medicine, this problem becomes acute. The late fourth and third centuries have many famous names in these fields, Diocles of Carystus, Praxagoras of Cos, Chrysippus of Cnidus, Herophilus of Chalcedon, Erasistratus of Ceos, but not a single complete treatise of any of these writers has survived. This would seem to make the task of assessing their work impossible. However, since later medical authors, Celsus, Rufus, Soranus and Galen, quote and comment on their predecessors, often at considerable length, the situation is not as hopeless as it appears at first sight. Galen is our most valuable source. Although he wrote more than four hundred years after the great Hellenistic biologists, he was well acquainted with their work, for which he often expresses his admiration and to which he was greatly indebted. The second-hand evidence for some parts, at least, of the work of the two most important biologists, Herophilus and Erasistratus, is substantial, although no adequate modern edition of the surviving quotations from their writings has yet been made. Both men worked at Alexandria in the first half of the third century B.C., Herophilus being the older of the two. Both were trained as doctors, but both had other interests outside medicine proper.

The Alexandrian biologists were responsible for one major, if shortlived, success: they were the first to practise dissection on the human body. Before them, such dissections as had been carried out were all performed on animals. Neither the Hippocratic writers—who refer only seldom to dissection—nor Aristotle—who does so much more frequently—had dissected the human body. Several of our sources speak of Herophilus and Erasistratus undertaking not merely dissections, but also vivisections, on human subjects. Admittedly some of the evidence is untrustworthy. When the Christian writer Tertullian

(c. 200 A.D.) describes Herophilus as 'that doctor or butcher who cut up innumerable corpses in order to investigate nature and who hated mankind for the sake of knowledge' (*On the Soul*, ch 10), his testimony by itself carries little weight. Tertullian was totally opposed to the scientific investigations of pagan researchers and did everything he could to defame them and their work.

We have, however, more reliable authorites. Pliny and Rufus both refer in general terms to the practice of human dissection without specifying who first undertook this. But another first century A.D. source, the Roman medical writer Celsus, both identifies the men concerned and reports the arguments that were used to justify this practice and that of vivisection. In the introduction (23 ff) of his work *On Medicine* Celsus writes as follows concerning the group of doctors known as the Dogmatists:

> Moreover since pains and various kinds of diseases arise in the internal parts, they hold that no one who is ignorant about those parts themselves can apply remedies to them. Therefore it is necessary to cut open the bodies of dead men and to examine their viscera and intestines. Herophilus and Erasistratus proceeded in by far the best way: they cut open living men—criminals they obtained out of prison from the kings—and they observed, while their subjects still breathed, parts that nature had previously hidden, their position, colour, shape, size, arrangement, hardness, softness, smoothness, points of contact, and finally the processes and recesses of each and whether any part is inserted into another or receives the part of another into itself.

The Dogmatists wrote of the advantages of vivisection over dissection and defended this viewpoint against the charge of inhumanity by claiming that the good outweighed the evil: 'nor is it cruel, as most people state, to seek remedies for multitudes of innocent men of all future ages by means of the sacrifice of only a small number of criminals.'

Unlike Tertullian, Celsus cannot be accused of malicious distortion. He himself disagrees with the Dogmatists. 'To cut open the bodies of living men,' he says later in his introduction (74 f), 'is both cruel and superfluous: to cut open the bodies of the dead is necessary for medical students. For they ought to know the position and arrangement of parts—which the dead body exhibits better than a wounded living subject. As for the rest, which can only be learnt from the living, experience itself will demonstrate it rather more slowly, but much more mildly, in the course of treating the wounded.' The tone of his whole account is restrained and we have no good grounds for rejecting it. No one can doubt that religious and moral considerations inhibited the opening of the human body, whether dead or alive, in antiquity. But that is not to say that such inhibitions could never, under any circumstances, be overcome. The situation at Alexandria in the third century B.C. was clearly an exceptional one in the particular combination of ambitious scientists and patrons of science that existed there at that time. For all the ancients' respect for the dead, corpses were desecrated often enough by people other than scientists. Moreover, when we reflect that the ancients regularly tortured slaves in public in the law courts in order to extract evidence from them, and that Galen, for example, records cases where new poisons were tried out on convicts to test their effects, it is not too difficult to believe that the Ptolemies permitted vivisection to be practised on condemned criminals.

Yet while we have no good reason to deny that Herophilus and Erasistratus carried out both dissections and vivisections on human subjects, the extent of their investigations is another matter. So far as we can tell, their knowledge of the internal anatomy of man was still quite limited. They made some notable discoveries, and Herophilus' accounts of the duodenum and the liver in particular have been taken to indicate definite first-hand knowledge of these organs in man. But they were still capable of making fairly simple mistakes—as for example when Herophilus is reported to have maintained—what several Greek anatomists believed—that the optic nerves are hollow.

Herophilus' chief work was in anatomy, on which he composed several treatises, including one *On Dissections* in several books, and where a number of the terms he coined passed, either directly or via their Latin translations, into anatomical vocabulary. Thus we know that he made a careful study of the brain which—against the view of Aristotle—he recognized as the centre of the nervous system. He distinguished the main ventricles of the brain and identified in them what he called the 'chorioid concatenations'—a 'plexus of veins and arteries held together by delicate membranes' (as Galen reports) which he named chorioid from the resemblance to the outer membrane of the foetus (chorion). He described the 'calamus scriptorius', the cavity in the floor of the fourth ventricle of the brain that derives its name from his having compared it with the groove of a writing pen, and in his account of the blood-vessels of the brain he identified the confluence of the sinuses which he called the 'winepress' (*lēnos*) and which later anatomists called, after him, the torcular Herophili. He dissected the eye and distinguished its principal membranes: his comparison with a net was the origin of the Greek term 'net-like' ('retiform') for the membrane that is still known as the retina. According to Galen, Herophilus was one of the first to undertake an extensive—even though still very incomplete—study of the nerves, and indeed he and Erasistratus were among the first to begin to distinguish clearly between sensory and motor nerves, and between these and other tissues, such as tendons and ligaments, that were also called *neura* in Greek. We may infer that Herophilus described the main chambers of the heart and the blood-vessels communicating with it from other passages in Galen and Rufus. These tell us that he treated the auricular appendages as independent compartments of the heart, separate from the atria, and that he coined the name 'arterial vein' for what we call the pulmonary artery—a term that continued in use right down to the time of Harvey. Yet another example of his successful coinage of anatomical terms is duodenum, the Latin translation of his Greek name *dōdekadaktylon*, derived from the length (twelve finger's breadths) of the duodenum in man.

In many cases little trace of his work remains except the term that he coined for the structures he identified. But in one of the longer fragments of *On Dissections* quoted by Galen (*On Anatomical Procedures* VI ch 8) we find him touching on problems in comparative anatomy. After briefly describing the liver in man, he goes on:

> The liver is not alike in all, but differs in different animals in breadth, length, thickness, height and number of lobes,

> and also in the irregularity of the front part at which it is thickest, and the arched top parts where it is thinnest. In some the liver does not have lobes at all but is round and undifferentiated. In some however it has two, in some more, and in many four lobes.[1]

His work on the reproductive organs also stands out for his discovery of the ovaries, whose structure and function he compared with those of the testes in the male—an analogy that proved fruitful even though it also led him to some mistaken conclusions.

Herophilus adopted a humoral pathology from earlier Greek medical writers, and he wrote on dietetics and pharmacology—some of his prescriptions have been preserved. But his most important contribution to the clinical medicine of his day was undoubtedly his development of the theory of the diagnostic value of the pulse. Although the pulse is referred to occasionally by earlier writers (for example by Aristotle in his *Inquiry Concerning Animals* 521 a 6f), it was Herophilus' teacher, Praxagoras, who first restricted the pulse to a distinct group of vessels and held that it could be used as an indicator of disease. Herophilus corrected his master's teaching on several points, maintaining that the pulse is not an innate faculty of the arteries, but one they derive from the heart, and distinguishing the pulse not merely quantitatively but also qualitatively from palpitations, tremors and spasms, which are muscular in origin. Above all he and his followers undertook a systematic classification of different types of pulse according to 'magnitude' 'speed' 'intensity' 'rhythm' 'evenness' and 'regularity'. He clearly understood the differences in the pulse rate that depend on age, and he identified three main types of abnormal pulse ('pararrhythmic' 'heterorrhythmic' and 'ekrhythmic') differing in the extent to which they diverge from the norm, as well as other special kinds of abnormal pulse, such as those he described as the 'ant-like' (*myrmēkizōn*) and the 'gazelle-like' (*dorkadizōn*).

When we reflect that Herophilus had no accurate means of timing the pulse-rate, his attempt to develop a systematic

[1] Based on the translation in C. Singer, *Galen, On Anatomical Procedures*, Oxford University Press, 1956.

theory of the pulse is astonishing. Much of the stimulus to this work came from musical theory. As Galen (K IX 464) reports: 'as the musicians establish their rhythms according to certain definite arrangements of time-periods, comparing the *arsis* and *thesis* with one another [that is, the upward and the downward beat], so Herophilus supposes that the dilatation of the artery corresponds to *arsis* and its contraction to *thesis*.' Herophilus' attempt to reduce the data concerning the pulse to mathematically expressible relations analogous to musical theory was doomed to failure. But if later clinicians ignored many of his fine distinctions, his insistence on the importance of the pulse in diagnosis was of lasting value.

Like Herophilus, Erasistratus was first and foremost a doctor. Fragmentary though our information is, it suggests a remarkable blend of interests. He seems to have been a cautious clinical practitioner, criticizing the use of drastic remedies such as blood-letting and strong purgatives common in Greek medicine. At the same time he put forward a bold and original physiological and pathological doctrine.

The most prominent feature of this is the use of what we should call mechanical ideas to explain organic processes. One instance is his account of digestion. Aristotle, for instance, had suggested that the food undergoes a qualitative change, 'concoction', in the stomach, a change that he held to be effected by the 'innate heat' in the body and that he compared with boiling. Erasistratus rejected the Aristotelian analogy, pointing out, among other things, that the heat involved in digestion is far less than that of boiling, and he attempted to explain the processes that go on in the alimentary canal as far as possible in mechanical terms. He knew that food is propelled along the alimentary tract by the peristalsis of the gullet and the contractions of the stomach. Food is, then, not attracted to the stomach (as Galen was to believe) but impelled along the alimentary canal by muscular action. Within the stomach it is submitted to further mechanical action, 'trituration' or pounding, and it is then squeezed out, in the form of chyle, through the walls of the stomach and the intestines into the blood-vessels communicating with the liver. Finally the nourishment is absorbed by the tissues by being drawn out through the walls of the blood-vessels. To explain this part of the process, he appealed to the principle of *horror vacui*, that is, the tendency of nature to fill a vacuum. He postulated the existence of a partial vacuum created in the tissues by the evacuation of certain residues, and he held that this vacuum causes some of the contents of the blood-vessels to be drawn out into the tissues themselves.

A second example of Erasistratus' use of mechanical explanations is his account of the secretion of residues such as bile and urine. Again he was criticized by Galen for neglecting the natural 'attractive' function of the liver and kidneys. But Erasistratus' theory referred to simple morphological factors. In the liver the unpurified blood passes through a complex of vessels of varying calibre, and it is because of the *relative size* of the vessels that the bile is separated out from the pure blood. Although Galen complains that he gave no clear explanation of the secretion of urine, it is likely that on this problem too he put forward a similar theory, in terms of the relative calibre of the vessels in the kidneys.

The most interesting part of Erasistratus' physiology is his account of the vascular system. First he clearly appreciated the difference between the two types of vessels, veins and arteries. In this he was, to be sure, far from entirely original. Aristotle and several of the Hippocratic writers were aware of certain general anatomical distinctions between the two, even though they mostly continued to use the term *phleps* ('blood-vessel') indiscriminately of both veins and arteries. In Erasistratus, however, there is a physiological distinction between the two types of vessels, for he took over, and elaborated, an idea that is also found in some Hippocratic texts, namely that while the veins contain blood, the arteries contain air.

In trying to understand why many Greek doctors held this, we must note, first, that the term *artēriā* was applied, orginally, not only to the major arterial vessels connecting with the heart, but also and more especially to the principal ducts of the respiratory tract, the wind-pipe and the bronchi. Our word trachea comes from the Greek name for that duct, *hē trācheia artēriā*, literally 'the rough artery', which continued to be used long after a clear distinction between the bronchi and what we call the arteries had been drawn. Secondly, most anatomical investigations were carried out on dead animals, and in these the blood naturally drains away from the arterial into the venous system. Thirdly, the marked difference in the colour (and pressure) of arterial and venous blood was also used to suggest that the contents of the two types of vessels are different.

Many theorists who recognized this difference explained it by supposing that whereas the veins contained blood alone, the arteries were filled by blood and air (*pneuma*). But other physiologists, including Erasistratus, took the view that in their normal state, the arteries contain nothing but air. It is, however, perfectly clear from Galen that Erasistratus knew very well that blood flows when an artery is cut. But this was a lesion, and he believed he could explain what happens on the basis of the general mechanical principles that he applied elsewhere in his physiology. The arterial system contains air, but when an artery is cut, this air escapes, creating a partial vacuum which draws blood into the arteries from the adjoining veins.

But although he adhered to this quite mistaken view concerning the contents of the arteries, he had a clearer understanding of other features of the vascular system than any earlier—and indeed than many later—Greek theorists. First he appreciated the role of the four main valves of the heart: indeed he was probably the first person to do so.[1] Galen's extensive discussions of this part of Erasistratus' work make it plain that he had a detailed knowledge of both their form and function. In particular, he understood that each acts as a one-way valve. Thus the tricuspid valve at the entrance of the right ventricle allows the blood to enter, but not to relapse into the right atrium and vena cava. At the base of the 'arterial vein' (that is, the pulmonary artery) there is a second, semilunar valve (the pulmonary valve) that allows blood to flow out towards the lungs, but not back into the heart. Similarly there are two valves—the mitral and aortic—that control the flow into and out of the left side of the heart, though here, in

[1] The treatise called *On the Heart* (chs 10 and 12) describes the semilunar valves at the base of the aorta and of the pulmonary artery and may also refer (though the text is obscure) to the atrio-ventricular valves. But although this work is included in the Hippocratic Corpus its date is far from certain and it may well have been composed after Erasistratus' investigations.

Erasistratus' view, it is air, not blood, that enters and leaves. Secondly, he appreciated that the heart acts as a pump and is directly responsible for the dilatation of the arteries. He compared the heart to a bellows and the arteries to sacks or bags into which air is pumped, and he correctly held that while the heart fills because it is dilated, the arteries dilate because they are filled. Indeed a text in Galen (*On Anatomical Procedures* VII ch 16) is evidence that Erasistratus established this thesis experimentally by inserting a tube into an exposed artery and observing that the pulse continued in the artery below (distal to) the tube, an experiment that Galen was to repeat himself, only to obtain the opposite result and claim that the artery below the tube was pulseless. Thirdly, Erasistratus was led to infer that there must be passages ('*anastomōseis*' or '*synanastomōseis*') communicating between the ends of the arteries and the veins. But although thus far his guess was correct, his idea of the function of these passages was quite different from that which we ascribe to the capillaries, and any suggestion that he arrived at the notion of the circulation of the blood would be quite misplaced. In his view, blood flows through the *anastomōseis* only in abnormal circumstances, when an artery is cut or some other lesion has been sustained, and when it does so, the flow is from vein to artery (not from artery to vein). The idea of the *anastomōseis* was based, in part, on anatomical observation: but they are essential to his theory, given that he believed that the arteries normally contain only air, but that nevertheless blood flows when one is cut.

Gaps remain in our evidence and (it seems) in Erasistratus' theories, but what emerges from the scattered fragments of his work is, for its time, a remarkably comprehensive physiology. Although Galen complained that he did not explain how or where the blood itself was produced, it is clear that he held that the venous system transports the end-products of the processes of digestion to every part of the body. The blood is carried up to the right heart by the vena cava and some is then pumped to the lungs through the pulmonary arteries, whose main function he probably saw as being to supply this blood as nourishment to the lungs. The respiratory, arterial and nervous systems all depend on air and are independent of the venous sytem. He knew that atmospheric air is drawn into the lungs through the windpipe and bronchi when the chest expands. But he also held (erroneously) that it is air that is drawn into the left ventricle of the heart through the pulmonary veins at each diastole, this air being then pumped out, at each systole, to the arteries all over the body. Air in some form[1] is also responsible, it seems, for the function of the nerves, which like the arteries and the veins, subdivide beyond the limits of our perception. Indeed he conjectured that a sub-sensible, three-fold complex of vein, artery and nerve is the basic element of which every tissue and organ of the body is composed.

The abnormal functioning of the body—like its normal processes—is also explained as far as possible in simple mechanical terms, the principle of *horror vacui* being, once again, especially important. One example is the case already mentioned of blood spilling out of the veins into the arteries. This may happen when a lesion occurs, or when the blood becomes too abundant in the body as a result of an imbalance of food and exercise. When the blood spilling over into the arteries meets the *pneuma* pumped out from the heart, this brings about a state of repletion, *plēthōrā*, which causes inflammation and fever. Other more specific conditions have other contributory causes, but morphological factors are again appealed to in, for instance, the account of dropsy, which was explained as the result of the hardening, and consequent narrowing, of the passages in the liver.

Erasistratus' work is an extraordinary combination of careful observation and bold, sometimes wild, speculation. Our sources associate him with the school of Aristotle, and it is easy to conjecture that his use of the principle of *horror vacui* is influenced by the work of Strato of Lampsacus on the vacuum (see above, pp 17 f), although elsewhere Erasistratus rejected

[1] Texts in Galen ascribe to Erasistratus a distinction between the 'vital spirit' (*pneuma zōtikon*), which is contained in the arteries and left ventricle of the heart, and the 'psychical spirit' (*pneuma psȳchikon*), responsible for the nervous system, which is elaborated from the 'vital spirit' and is located chiefly in the brain. Both types of *pneuma* are ultimately derived from the air we breathe, but how this is modified to become the two types of *pneuma* is not clear from our information: nor is it clear how far Galen may have assimilated Erasistratus' ideas to his own, in some ways similar, theory (see below, pp 140 f).

such key Aristotelian doctrines as that of the primary importance of the 'innate heat' in the natural processes of the body. Galen criticizes him for not being clear on the ultimate constitution of matter itself, but so far as organic tissue goes, the notion of the basic three-fold element of nerve-vein-artery adds a new idea to the debate that stretches back into the Presocratic period, in which doctors and philosophers alike had speculated on 'the nature of man' and the ultimate constituents of the human body. Yet while he frequently conjectures about what is beyond the limits of perception, some of his theories are backed by a greater use of observation than we find in most Greek writers. Much as he is criticized by Galen as a physiologist, as a descriptive anatomist he won his praise not only for his account of the valves of the heart, but also for his work on the brain and the nervous system. Moreover there is some evidence that he attempted to support some of his theories by means of deliberate tests. One example of this, the experiment of inserting a tube into an artery to investigate the pulse, has already been mentioned (p 83). Another comes from the history of medicine by the writer known as Anonymus Londinensis (XXXIII, 43 ff). According to this he adopted from earlier theorists the doctrine that invisible effluences are emitted from the body and apparently tried to establish this by a test on a living bird. The bird was first weighed and then kept in an enclosed space for a period of time. It was then reweighed together with the excrement it had passed and the total weight was found to be less than at the beginning—a fact that Erasistratus interpreted as showing, as our source puts it, that 'a considerable emanation had taken place'.

Herophilus and Erasistratus represent the high-water mark of Alexandrian biology. They had the imagination to see the need to practise dissection on the human as well as the animal body, and they had the good fortune to have the support of the Ptolemies in this. We have only to compare the Alexandrians' work on the eye and the heart with earlier accounts in Aristotle and the Hippocratic or Presocratic writers to realize the advance that anatomy made in a short space of time in Alexandria. The problems of physiology did not yield to solution so readily as some of the anatomical questions they investigated. Yet Erasistratus, in particular, had some ingenious ideas to contribute even if—as the example of his theory of the contents of the arteries shows—in putting forward some of his conjectures he was obliged, and prepared, to explain some well-known data away. Yet interestingly enough Erasistratus himself provides us with one of the classic statements, in Greek science, of the determination and persistence that scientific research demanded. This is in a text from the second book of his work *On Paralysis*, for which our source is, once again, Galen: (*On Habits* ch 1):

> Those who are completely unused to inquiry are, in the first exercise of their mind, blinded and dazed and straightway leave off the inquiry from mental fatigue and an incapacity that is no less than that of those who enter races without being used to them. But the man who is used to inquiry tries every possible loophole as he conducts his search and turns in every direction and so far from giving up the inquiry in the space of a day, does not cease his search throughout his life. Directing his attention to one idea after another that is germane to what is being investigated, he presses on until he arrives at his goal.

The history of the biological sciences in the generations that immediately followed Herophilus and Erasistratus is obscure, but it is clear that the practice of human dissection declined, even if it did not completely die out, and we must ask why this was so. Our main evidence comes from Rufus and Galen. A passage in Rufus (end of first century A.D.) shows that although he considered human dissection the ideal, this was a thing of the past: 'We shall try to teach you how to name the internal parts by dissecting an animal that most closely resembles man . . . In the past they used to teach this, more correctly, on man' (*On the naming of the parts of man*, 134). But two texts in Galen's *On Anatomical Procedures* provide fuller information which modifies the picture somewhat. In the first (I ch 2) he refers to the study of the bones:

> Make it rather your serious endeavour not only to acquire accurate book-knowledge of the shape of each bone but also to examine assiduously with your own eyes the human bones themselves. This is quite easy at Alexandria because

> the physicians there employ ocular demonstration in teaching osteology to their students. For this reason, if for no other, try to visit Alexandria.[1]

This indicates that in one context (the study of the bones) and at one place (Alexandria) human dissection continued to be practised in Galen's own day (second century A.D.). For other purposes, and in other parts of the Greco-Roman world, this was much more difficult. Galen continues in the same passage:

> But if you cannot [visit Alexandria], it is still possible to see something of human bones. I, at least, have done so often on the breaking open of a grave or tomb. Thus once a river, inundating a recent hastily made grave, easily broke it up, and completely swept away the body of the dead man with the force of its movement. The flesh had putrefied, though the bones still held together in their proper relations . . . This skeleton was as though a doctor had deliberately prepared it for such elementary teaching. And on another occasion we saw the skeleton of a brigand, lying on rising ground a little off the road. He had been killed by some traveller repelling his attack. None of the inhabitants would bury him, but in their hatred of him were glad enough to see his body consumed by the birds which, in a couple of days, ate his flesh, leaving the skeleton as if for demonstration.

Our second text comes from book III ch 5 of the same treatise, when, dealing with the study of the blood-vessels, he writes of the need for frequent practice in animal dissection,

> so that if you have the luck to dissect a human body, you will be able readily to lay bare each of the parts. This is not everybody's luck, and it cannot be achieved at short notice by one unskilled in the work. Even the greatest experts in anatomy among the physicians, and even when examining the parts of the body at leisure, have obviously made many mistakes. For such a reason even those who sought to dissect the body of a German enemy, who had been killed in the war against Marcus Antoninus, could learn no more than

[1] Based on the translation of C. Singer, *Galen, On Anatomical Procedures*, Oxford University Press, 1956.

the position of the viscera. But one who has practised beforehand on animals, and especially on apes, lays bare with the utmost ease each of the parts for dissection.

Both Rufus and Galen held that the correct method was to practise dissection on the human body, and Galen at least had some actual experience of this. Yet other writers contested this view. Some were sceptical about the value of dissection in general, insisting on the difference between the dead body and the living and maintaining that the physician, whose task was to heal the living, could learn nothing useful from the study of the dead. Others advocated dissection but thought that animal dissection was adequate for their purposes. Although in a moral or religious context many ancient writers emphasized the difference between the souls of men and animals, for the sake of studying man's *body* it was commonly held that animals, particularly apes or other species close to man, could be used as a guide.

Human dissection had never been generally accepted as an *essential* part of medical training, let alone of research. The doctors themselves disagreed about its value. Those who, like Galen, recognized its usefulness were faced with formidable practical difficulties, not least that of obtaining a body. It is here that the contrast between Galen's situation and that of Herophilus and Erasistratus is most marked. The story of the German soldier shows that the authorities occasionally made bodies available for dissection in the second century A.D. But otherwise, as our other passage from Galen illustrates, the doctor was reduced to taking such chance opportunities as presented themselves. In these circumstances even those who advocated human dissection as the ideal method took the easy way out and used animal subjects for most of their work.

The detailed history of anatomy and the other special branches of medicine cannot be followed up here. But two general features of medicine after Erasistratus must be mentioned very briefly in conclusion, namely the proliferation of medical sects and the continuing interaction of medicine and philosophy. Some of these sects were named after individuals, including Herophilus and Erasistratus themselves: there were doctors known as Erasistrateans in the second century A.D. Other groups took their name from the views they held in the complex and protracted controversy on the correct method in medicine. Two of the main schools were the Dogmatists and the Empiricists.[1] Against the Dogmatists, who argued that knowledge of 'hidden causes'—in particular of the constitution of man and of the causes of diseases—is essential to medical practice and that such knowledge can only be obtained by supplementing experience with reasoning and conjecture, the Empiricists asserted that it is neither legitimate nor necessary to speculate on such matters. For the Empiricist, the invisible cannot be known: the doctor's task is to treat individual cases and for this purpose he must avoid inference and attend to, and be directly guided by, the manifest symptoms of the patient and these alone.

This controversy reflects and was no doubt directly influenced by contemporary philosophical debates on the foundation of knowledge, although both view-points also owe much to earlier medical writers. Against the Peripatetics, Stoics and Epicureans, all of whom maintained the possibility of knowledge, though they gave different accounts of its basis, different types of sceptical philosophy were put forward, first by Pyrrho of Elis (fourth century), then in the Academy under Arcesilaus (third century) and then by Aenesidemus and his followers (first century B.C.). Some sceptics denied that knowledge is possible: others took the view that such a denial was itself a dogmatic assertion and that the sceptic must on this, as on every other, matter withhold judgement: but all agreed in rejecting every attempt to establish a definite criterion of

[1] A third sect, known as the Methodists, became fashionable at Rome in the first and second centuries A.D., the most outstanding representative being Soranus of Ephesus who wrote important works on gynecology and pathology. The origins of the sect are obscure: its doctrines were traced back to Themison in the first century B.C., though it may be that we should consider Thessalus, at the beginning of the next century, the founder of the sect. Whereas the Empiricists adopted an epistemology close to that of the Academic Sceptics who asserted that the non-evident cannot be grasped, the position of the Methodists corresponded to that of the later Sceptics who maintained that judgement should be withheld (see, for example, Sextus Empiricus, *Outlines of Pyrrhonism*, I 236–241).

knowledge. But while general epistemological arguments were taken over by the medical writers in their debates concerning the aims, methods, nature and justification of medicine, they also adduced considerations derived specifically from their own medical experience, and on occasions the Empiricists, especially, explicitly dissociated themselves from the philosophers. This can be illustrated from Celsus' report concerning their views (*On Medicine*, introduction 27 ff). They argued that nature cannot be comprehended partly on the grounds of the disagreements concerning causes among both philosophers and medical practitioners. 'Even students of philosophy would have become the greatest medical practitioners, if reasoning could have made them so; but as it is, they have words in plenty, but no knowledge of healing at all.'

The theme of the distinction between theory and practice, between words and action, so common in so many contexts in Greek thought, goes back, in medicine, to our earliest texts, the Hippocratic treatises. When it reappears in the methodological controversies of late Hellenistic medicine, it is sometimes combined with a sceptical view-point. With the Empiricists, medicine may be said to become the antithesis of mathematics. The latter is the theoretical inquiry par excellence, and the model of certain knowledge. Medicine provided the sceptic with many of his chief examples and arguments and at least one notable sceptical philosopher of the second century A.D., Sextus Empiricus, was by training also a doctor. At the same time, the Empiricists' rejection of doctrine and inference reflects their views of the practical aims of medicine. Speculation is not only illegitimate, but also, they believed, superfluous, for to heal the sick the doctor need have no recourse to general theories.

7

Applied Mechanics and Technology

THE investigation of ancient attitudes towards the possibility of applying scientific knowledge to practical ends, like any question of a similar degree of generality, is fraught with difficulty and carries with it a temptation to over-simplification. Nevertheless we have texts that help to throw light on *some* ancient attitudes on this matter. The theme of the practical utility of technical knowledge is expressed, in certain contexts, in the sixth and fifth, as well as fourth, centuries B.C.[1] After Aristotle, too, the distinction between theoretical and practical inquiries is often explicitly drawn and different writers express their views on how knowledge may be put to practical use. What is interesting in the passages in question is first what is included under the heading of mechanics, secondly the conception of the 'usefulness' of knowledge, and thirdly the relative evaluations of what we should describe as the 'pure' and the 'applied' departments of scientific inquiry.

One of our fullest texts is in the *Mathematical Collection* (VIII, 1–2) of Pappus of Alexandria (early fourth century A.D.).

> The study of mechanics ... being useful for many important things in life, is with reason thought by philosophers to be worthy of the highest approval and is eagerly pursued by all those interested in mathematics ...
>
> The mechanicians associated with Hero say that mechanics has a theoretical and a practical part. The theoretical part consists of geometry, arithmetic, astronomy and physics, the practical part of metal-working, building, carpentry, painting and the manual activities connected with them. One who has been brought up in those branches of knowledge from childhood and has acquired skill in those arts and who has a versatile nature will, they say, be the best inventor of mechanical devices and the best master-craftsman

[1] See *Early Greek Science*, pp 133 ff.

(*architektōn*). But where it is not possible for the same man to excel in so many branches of mathematics and at the same time to learn the arts we have mentioned, they recommend a person who wishes to engage in mechanical work to make use of those particular arts he has mastered for the ends for which each is useful.

The most necessary of the mechanical arts from the point of view of the needs of life are these. [1] The art of the constructers of pulleys, who were called mechanicians by the ancients. With their machines they use a lesser force to raise high great weights against their natural tendency. [2] The art of the makers of instruments necessary for war, for these too are called mechanicians. Missiles of stone, iron and the like are hurled a great distance by the catapults that they make. [3] The art of the machine-makers properly so-called. Water is easily raised from a great depth by the water-lifting machines that they construct. [4] The ancients also called the wonder-workers mechanicians. Some invented pneumatic devices, as Hero in his *Pneumatics*, others seem to imitate the movements of living things by means of sinews and ropes, as Hero does in his *Automata* and *On Balances*, and others use floating objects, as Archimedes in his work *On Floating Bodies*,[1] or water-clocks, as Hero in his work *On Water-Clocks*, which is evidently connected with the study of the gnomon [sun-dial]. [5] They also call mechanicians those who are skilled in sphere-making, who construct a model of the heavens by means of the uniform circular motion of water.

A similar tradition underlies the account found in Proclus (*Commentary on the first book of Euclid's Elements*, 41 3 ff) who draws particularly on the work of the first century B.C. mathematician and astronomer Geminus. In Proclus mechanics is defined as that part of the 'study of material objects perceived by the senses' that embraces (1) the manufacture of engines useful in war, (2) the manufacture of wonderful

[1] In view of the nature of Archimedes' book—an abstract, mathematical discussion of hydrostatical problems, see above, pp 47 f—it is surprising that it should be mentioned here to illustrate the 'wonder-working' branch of mechanics.

devices, based on air currents, weights or ropes, (3) the study of equilibrium and centres of gravity, (4) sphere construction and (5) 'in general, the whole subject of the movement of material bodies'.

One striking feature of both these lists is the prominence given to the design and construction of weapons of war, and this was, in fact, one of the chief areas in which mechanical ideas were applied and developed in the Hellenistic period. A second, more surprising, item is the mention of 'the manufacture of wonderful devices', for example to simulate the movements of living things. When Pappus includes this as one of the 'most necessary' branches of mechanics from the point of view of practical utility, his conception of the latter is, clearly, a wide one: the 'usefulness' of the devices in question lies in their entertainment value or in their being employed to produce 'miraculous' effects in the service of religion. Again both writers mention as a separate branch of mechanics the construction of spheres to represent the movements of the heavenly bodies—such as we know that Archimedes made (see above, p 41). 'Practical utility' evidently also includes devices of use in the study of astronomy.

The passages we have mentioned so far contrast theoretical and practical inquiries without expressing any view on the relative value of the two types of investigation. But as is well known there are plenty of occasions when ancient authors do just that. One classic text in which a writer expresses a forthright opinion on this subject is in Plutarch. In his *Life of Marcellus* he describes how the mechanical inventions of Archimedes kept the Roman army at bay in the siege of Syracuse in 212 B.C., and he comments on the two aspects, theoretical and practical, of Archimedes' genius. According to Plutarch (ch 14),

> [Archimedes] by no means devoted himself [to the construction of engines] as work worthy of serious effort, but most of them were mere accessories of a geometry practised for amusement, since in bygone days Hiero the king had eagerly desired and at last persuaded him to turn his art somewhat from abstract notions to material things, and by applying his reasoning somehow to the needs which make

themselves felt, to render it more evident to the common mind.[1]

Plutarch goes on to give a sketch of the early history of mechanics, which he says was originated by Eudoxus and Archytas, whose work drew down on them the wrath of Plato:

> [Plato] inveighed against them as corrupters and destroyers of the pure excellence of geometry, which thus turned her back upon the incorporeal things of abstract thought and descended to the things of sense, making use, moreover, of objects which required such mean and manual labour. For this reason mechanics was made entirely distinct from geometry, and being for a long time ignored by philosophers, came to be regarded as one of the military arts.

After recounting some of Archimedes' feats of engineering, Plutarch concludes (ch 17):

> And yet Archimedes possessed such a lofty spirit, so profound a soul, and such a wealth of theoretical insight, that although his inventions had won for him a name and fame for superhuman sagacity, he would not consent to leave behind him any treatise on this subject, but regarding the work of an engineer and every art that ministers to the needs of life as ignoble and vulgar, he devoted his earnest efforts only to those studies the subtlety and charm of which are not affected by the claims of necessity . . . And although he made many excellent discoveries, he is said to have asked his kinsmen and friends to place over the grave where he should be buried a cylinder enclosing a sphere, with an inscription giving the proportion by which the containing solid exceeds the contained.

In considering this passage we must first recognize that Plutarch does not quote Archimedes directly, but rather ascribes certain views to him. Plutarch himself was no engineer,

[1] Based on the Loeb translation of B. Perrin, *Plutarch's Lives* vol 5, Cambridge, Mass., Harvard University Press; London, Heinemann, 1917.

but a well-to-do country gentleman, a man of letters with an interest in history and philosophy, where he was sympathetic to Platonism. When he implies that Archimedes valued his work in mathematics above all else, this rings true to life. The story of Archimedes' instructions concerning the monument on his grave may be apocryphal, but when we consider his extraordinary achievements in mathematics (see above, Chapter 4), it is in no way surprising that he should have wished to be remembered chiefly for them. On the other hand, when Plutarch also suggests that Archimedes positively despised 'every art that ministers to the needs of life', we may be more sceptical and suspect that here Plutarch's own Platonic bias obtrudes. The reports that associate Archimedes with such mechanical devices as the screw named after him (a device for lifting water which he is said to have invented on a visit to Egypt) and the compound pulley (which he is supposed to have demonstrated by drawing to himself, single-handed, a fully laden ship) may well have been embroidered by the tradition. Yet these reports definitely suggest that Archimedes was interested in mechanical problems from a practical, as well as from a theoretical, point of view.

Plutarch has fairly clearly exaggerated, if indeed he has not completely fabricated, Archimedes' distaste for engineering. Yet it is still significant that we find these views expressed in the *Life of Marcellus*, even if they are Plutarch's own more than Archimedes'. The educated elite whom Plutarch typifies generally combined contempt for the life of the engineer with ignorance concerning his work. This attitude, which had the weighty support of Plato and Aristotle, is, without a doubt, the dominant one in writers of all periods in antiquity.

There is, however, another side to this question. Plutarch had no practical experience of mechanics himself. Yet apart from those historians, philosophers or men of letters who chose to comment on the work of the engineers, treatises on aspects of mechanics were occasionally written by men who had at least some direct experience in this field. Four names, especially, should be mentioned, Ctesibius of Alexandria (who was working about 270 B.C.), Philo of Byzantium (c. 200 B.C.), Marcus Vitruvius Pollio (c. 25 B.C.) and Hero of Alexandria (c. 60 A.D.). Each of these men wrote on mechanical subjects, and although Ctesibius' book is lost, we have parts of Philo's so-called *Mechanical Collection*, Vitruvius' *On Architecture* and a number of Hero's works in Greek or in Arabic translation, for example *On Pneumatics*, *On Artillery Construction* and *On the Construction of Automata*.[1] These texts provide direct evidence concerning the engineers, the conditions under which they worked, the problems they investigated and the devices they were interested in.

Vitruvius, himself an architect-engineer employed by Augustus, is our most valuable source concerning the profession to which he belonged. The 'architectus' or *architektōn* might be responsible not only for the planning and construction of buildings or even whole towns, but also for the design, construction and maintenance of mechanical devices of various sorts, particularly weapons of war. Thus Vitruvius was employed to repair and rebuild the war-engines in the imperial army. Like the doctors, the architects were often anxious to establish the respectability of their calling. Vitruvius insists that an architect should have, as he had, a general education in such subjects as philosophy and mathematics, as well as a technical training. He is also keen to deny studying 'the art' for mercenary motives: 'rather have I held that a little money with a good reputation is to be pursued rather than much wealth with disgrace' (VI preface 5). The difficulties that the architect faced in finding employment whether from a rich individual or from a city come out vividly. The profession was highly competitive and Vitruvius writes of corruption in the awarding of commissions. The theme of the dependence of the architect on the patron is a recurrent one, and indeed one of the primary motives of the work *On Architecture* itself was to win the favour of Augustus.

Ancient writers on mechanics tackle a wide range of problems and describe a great variety of mechanical devices. How much any particular invention or technological development owed to any given theorist is usually impossible to determine, although a remarkable number of devices, including a pump, a water-clock and improvements to the catapult, are ascribed

[1] Hero also wrote extensively on geometry and mensuration, for example the *Metrica* and a commentary (of which only fragments survive) on Euclid's *Elements*.

by name to Ctesibius. In many cases, no doubt, the more important technological advances were achieved through the work of craftsmen who had little or no interest in theoretical issues. For their part, however, the writers on mechanics not only describe some complex devices but also discuss the mechanical principles involved, and they occasionally provide evidence of deliberate research undertaken to find the optimum solution to a practical problem.

One main area where technological advances were made through the application of mechanical principles was in warfare, where we can supplement the written texts with the evidence from the material remains, in particular representations of ancient weapons. The broad outlines of the history of the development of artillery in the ancient world have become clearer through the studies of E. W. Marsden who has shown how, beginning with modifications to the simple bow, more and more effective weapons were devised, particularly those incorporating the torsion principle to exploit the power of skeins of twisted hair or sinew (see Fig. 14). Reconstructions of some of the ancient weapons have enabled him to give estimates of the limits of their performances: thus he arrives at a figure of 400 yards for the maximum effective range of the catapult. Progress was comparatively rapid for some 150 years from the beginning of the fourth century (so far as we know, the first occasion when artillery was used at a siege was at Motya in 397) and improvements continued to be made down to the first century A.D. A text in Philo of Byzantium, to which I have already alluded (p 4), helps to explain part of this development. Philo tells us that the engineers were financed, in their investigations, by the Ptolemies and he also throws important light on the nature of those investigations themselves:

> Now some of the ancients discovered that the diameter of the bore [that is, the circle that receives the twisted skeins] was the basic element, principle and measurement in the construction of artillery. But it was necessary to determine this diameter not accidentally or haphazardly but by some definite method by which one could also determine the proportionate measurement for all magnitudes [on the instrument]. But this could not be done except by increasing

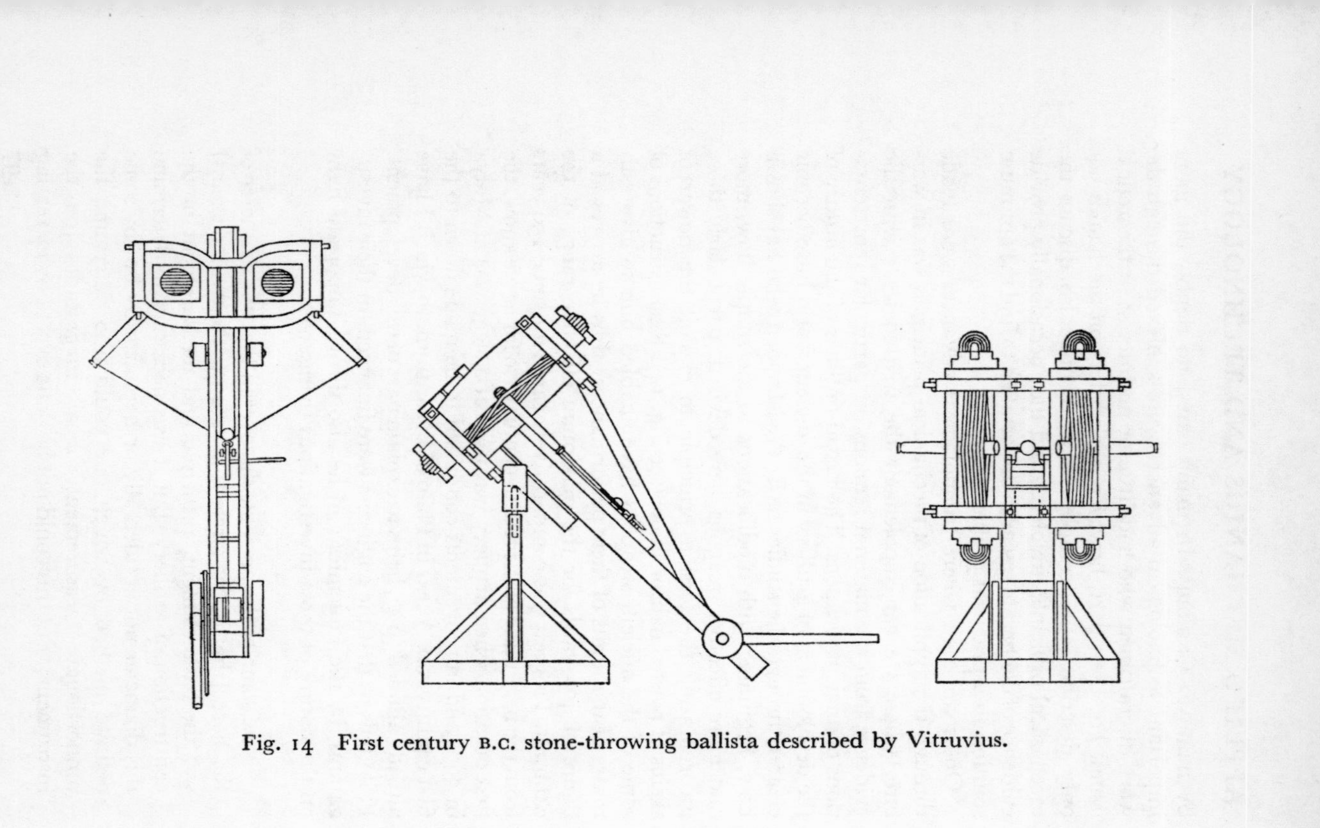

Fig. 14 First century B.C. stone-throwing ballista described by Vitruvius.

or decreasing the diameter of the bore and testing the result. And the ancients, as I say, did not succeed in determining this magnitude by test, because their trials were not conducted on the basis of many different types of performance, but merely in connection with the required performance. But the engineers who came later, noting the errors of their predecessors and the results of subsequent experiments, reduced the principle of construction to a single basic element, namely the diameter of the circle that receives the twisted skeins. Success in this work was recently achieved by the Alexandrian engineers, who received considerable support from kings who were eager for fame and well-disposed to the arts and crafts. For it is evident that it is not possible to arrive at a complete solution of the problems involved merely by reason and by the methods of mechanics, and that many discoveries can be made only as a result of trial. (*On Artillery Construction*, ch 3, 50 20 ff)[1]

This text shows quite clearly that the ancients did, on occasion, appreciate the need to conduct systematic tests in order to isolate the relevant variables and determine the relationship between them. Philo rejects *both* a simple trial and error approach *and* a priori dogmatism in favour of a method of controlled experiment. Testing for the required performance did not reveal the operative principles. Nor can the solution be obtained merely 'by reason and the methods of mechanics'.

For obvious reasons, military technology sometimes attracted considerable state support. But the ingenuity of Greek engineers was applied in other areas and to many other devices of varying usefulness and practicability. Of the five main simple machines known in the ancient world, four, namely the lever, the pulley, the wedge and the windlass, were in use long before the end of the fourth century B.C. But the fifth, the screw, was, so far as we know, an innovation of the third century. One of the first applications, if not the very first, has already been mentioned, the water-lifting device still

[1] Based on the translation in M. R. Cohen and I. E. Drabkin, *A Source Book in Greek Science* (second edition), Cambridge, Mass., Harvard University Press, 1958.

known as the Archimedean screw (see Fig. 15). Whether or not we accept the tradition that Archimedes himself was personally responsible for this invention, the screw, far more than any of the other simple machines, depends on working out and applying a mathematical construction. A second important application of it is in the screw-press. The simplest oil- or wine-press consisted of a lever or a beam on which pressure is exerted directly. This was first modified by various mechanisms, a rope passed over a drum, or a screw, to increase the pressure on the end of the lever. Then came the screw-press proper, in which pressure is exerted not indirectly on the end of a beam, but directly on the top of the press itself by a screw or a pair of screws. Pliny, writing in about 75 A.D., reports

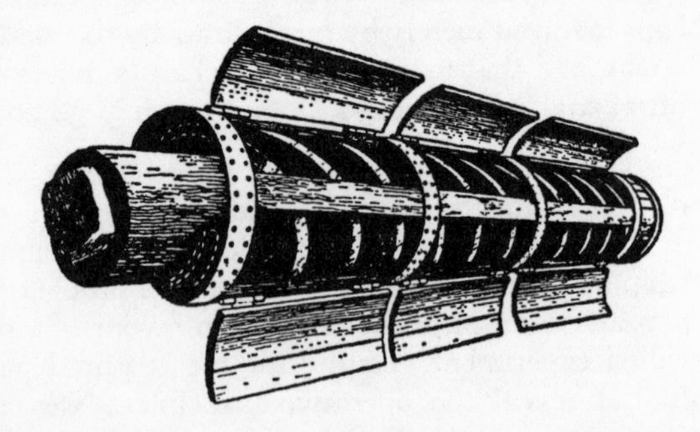

Fig. 15 Archimedean screw of oak from a mine at Sotiel in Spain.

(XVIII, 74, 317) that it was introduced 'within the last twenty-two years', although some sort of screw-press appears to be referred to in Vitruvius (c. 25 B.C., VI, 6, 3). A detailed account of a double screw-press is given by Hero (*On Mechanics* III ch 19, see Fig. 16), who besides using the screw in many of his gadgets provides our first extant description of a machine for cutting screws. Other useful devices in our mechanical texts include a series of other water-lifting machines, cranes, surveying instruments and clocks. Thus among the water-lifting devices described by Vitruvius is one that he ascribes to Ctesibius in particular (X 7): it is known as the 'fire-engine' from the fact that Hero refers to it as the 'siphon used in conflagrations'. This is a double force pump, incorporating a system of valves, cylinders and pistons (see Fig. 17). Whether or not it was ever actually used to put out a fire, archaeological evidence confirms the use of valves and plungers in ordinary water pumps in antiquity and we have no reason to doubt Ctesibius' interest in the principles involved. Vitruvius also tells us (IX 8 2 ff) that Ctesibius was one of the first to investigate the principles of construction of water-clocks: he ascribes to him a constant head water-clock and various devices by which the length of the hour—conceived by the ancients not

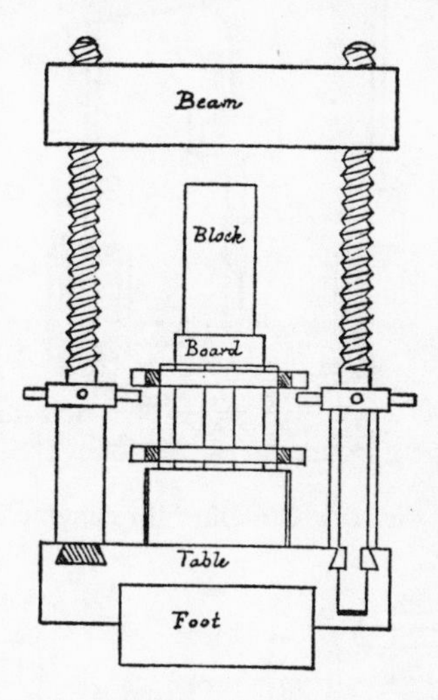

Fig. 16 Reconstruction of twin screw-press described by Hero *On Mechanics* III 19.

as an absolute unit of time, but as a division of a period of sunlight—could be adjusted according to the season of the year (see Fig. 18).

The devices we have been considering were all designed to serve practical purposes of one kind or another. But many other gadgets were invented purely for amusement. As we have seen, both Pappus and Proclus treat the contriving of wonderful devices as a separate branch of mechanics, and this corresponds to views expressed in our earlier mechanical texts. Thus in the introduction to his *Pneumatics* Hero distinguishes between arrangements that 'supply the most necessary wants of human life' and those that 'produce astonishment and wonder'. Many of his applications of pneumatic principles come in the latter category. He describes more than two dozen gadgets constructed with secret compartments and interconnecting pipes and siphons that enabled strange effects to be produced—Magic Drinking Horns from which two different

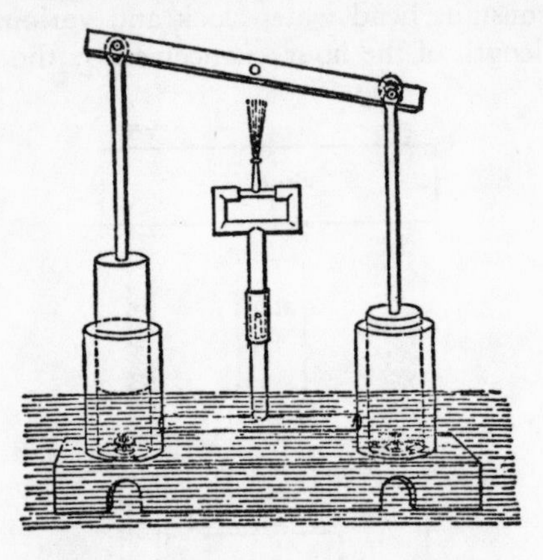

Fig. 17 Ctesibius' 'fire engine'.

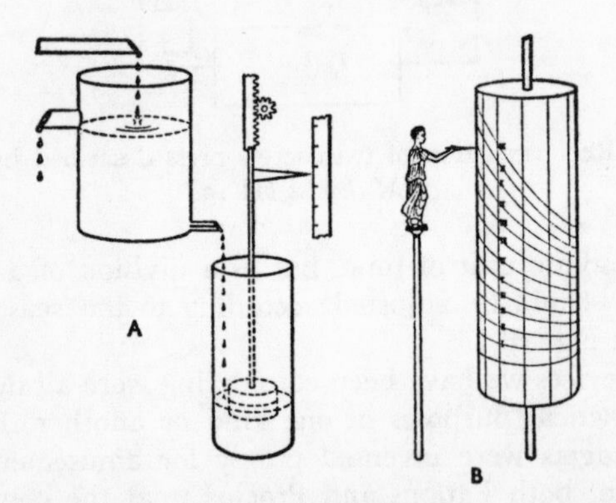

Fig. 18 Constant-head water-clock with simple pointer (A) and (B) one method of adjusting the length of the hours according to the seasons.

liquids can be poured, for instance, or Magic Mixing Vessels that replenish themselves (from a hidden reservoir) as they are emptied. Several of these gadgets are designed to be used in religious cults. One such device, which uses the expansion of air when heated, is described in *Pneumatics* I ch 12. In this figures standing on a hollow altar are made to pour libations when a fire is lit on the altar: the fire causes the air within the altar to expand and drive out the liquid contained in the altar pedestal and this liquid passes up tubes hidden in the figures' bodies and appears to be poured by the figures on the altar (Fig. 19). In an even more ambitious device (*Pneumatics* I

Fig. 19 Libations at an altar produced by fire.

ch 38) temple doors are made to open and shut automatically when a fire is lit: again the motive force is the expansion and contraction of air (Fig. 20). Other devices use the power of steam. This had been remarked on by Aristotle, for example, who observed that 'as a fluid turns to steam and vapour, the vessels that contain the substances burst for lack of room' (*On the Heavens* 305 b 14 ff). The best known gadget of this type is the toy described in Hero *Pneumatics* II ch 11. In this, a hollow ball with bent tubes attached is made to rotate on a pivot over a cauldron of water which is heated to make steam (Fig. 21).[1]

Finally the most elaborate toy of all was the 'automatic theatre', described in Hero's treatise of that name, in which a small stage comes into view, a puppet show representing for example the work of a shipyard is presented, and the stage

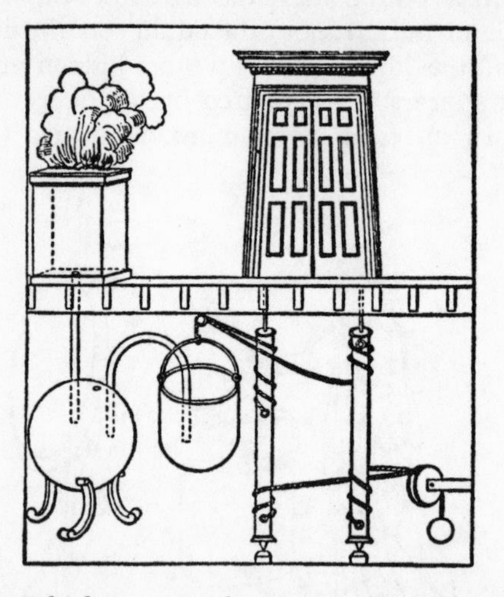

Fig. 20 Temple doors opened automatically by fire on an altar.

retires again—all automatically. A. G. Drachmann describes the mechanism as follows:

> The moving force is a heavy weight fitting into a container full of millet or mustard seeds; the seeds run out through a narrow hole, the weight comes down at a determined rate, and it turns an axle from which it is suspended by a cord. All the movements are taken from this axle by means of strings. A puppet or any other thing is turned by a string going round a drum; if it has to be turned back, the string

[1] This is sometimes referred to, quite misleadingly, as Hero's 'Steam Turbine'. The motion is certainly rotative, but what Hero describes is a toy, not a practical device for employing the power of steam (see below, p 106).

is passed over a peg in the drum and wound round the other way. If it has to move, and stop, and move again, there is a length of slack string between two windings ... A movement of the arm of a puppet, e.g. hammering, is produced by pins on a wheel acting on the short end of a lever.[1]

Studying the texts of ancient writers on mechanics, one is struck by three things especially, first the ingenuity with which they thought up new applications of a limited number of simple mechanical principles, second the interest they showed

Fig. 21 Hero's ball rotated by steam.

in those principles themselves, and in the theoretical side of mechanics as a whole, and third the fact that they are aware of, and distinguish between, two types of aim or justification for these studies, that is, to serve practical ends, and to amuse or amaze. Given that the ancient writers recognized the possibility of applying mechanics to practical needs, it is surprising that this did not lead to more fruitful results. Technology was far from being as stagnant, during the period from about 500 B.C. to about 500 A.D., as is sometimes suggested. Considerable advances were made not only in military technology but also in agriculture and food technology—as

[1] In *The Mechanical Technology of Greek and Roman Antiquity*, Copenhagen, Munksgaard, 1963, p 197.

for example L. A. Moritz's study of milling techniques has shown. The mechanical devices that were invented during that period include, besides the screw-press, the compound pulley, gear- and cog-wheels, the suction pump and the water-mill. Nevertheless the list of such devices is a modest one, and we are led to inquire why this is so, and why, in particular, the ancients were slow to exploit, or sometimes quite failed to exploit, mechanical principles that were known to them.

This general problem can only be tackled by taking some specific examples. A case that is often mentioned is the failure to exploit the power of steam. Hero, as we have noted, describes a ball that is made to rotate on pivots by steam escaping from bent tubes attached to it. Yet to claim, as has sometimes been done, that all the elements of a steam engine are already present, potentially, in this toy is absurd. The harnessing of steam depended, among other things, on being able to cast large metal cylinders accurately and effect clearances between piston and cylinder fine enough to prevent the escape of steam as pressure builds up, and on devising an efficient method of converting rectilinear to rotative motion. The problems that had to be overcome to make an efficient steam engine were formidable and it was only after a long and complex process of development that an engine capable of more than 10 horse power was finally produced in the late eighteenth century.

A more interesting and significant example is the slow diffusion of the water-wheel used as a source of power. From the first century B.C. this is mentioned by a number of authors and there is a brief description in Vitruvius (X 5 2) in particular (Fig. 22). Yet it appears that a good while elapsed before water-wheels came to be used at all commonly as a power-source. We have little definite archaeological evidence for ancient water-wheels in general and none of any wheel before the second century A.D. Systematic exploitation of the power of water as in the series of sixteen wheels at Barbegal near Arles had, it seems, to wait until even later: the mill at Barbegal was first used in the mid third century A.D. and by the end of the fourth century the use of the water supply to feed mills was the subject of legislation and litigation. Yet these developments were remarkably slow in coming.

One obstacle to the diffusion of the water-wheel that has often been mentioned is the lack of suitable water-supplies in many parts of the ancient civilized world. The ideal water-supply is a fast-flowing stream that is constant all the year round. While these are common enough north of the Alps, they are rare in Greece, Italy and Asia Minor. Even so, this problem could be met by supplying the wheel by an artificial aqueduct—and indeed all the main ancient water-wheels that have been excavated are so supplied. Again the problem could

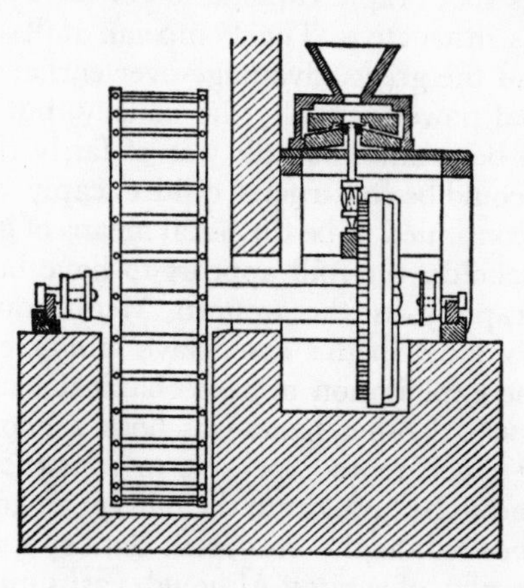

Fig. 22 Roman water-mill according to Vitruvius.

be solved in another way, by constructing the wheel on a floating platform anchored in the river itself, and this solution too was known in late antiquity. Thus Procopius (*On the Gothic War*, I 19, 19 ff) describes one such floating mill constructed in the Tiber by Belisarius during the siege of Rome in 537.

The problem of water-supply is clearly not the only factor that has to be considered to explain the slow diffusion of the water-wheel. It has often been argued that the key to this question—and to the backwardness of ancient technology as a whole—lies in the institution of slavery. So long as slaves were readily available, there was, so it is said, no incentive to devise artificial power-sources or labour-saving techniques of any sort. There is no doubt a good deal of truth in this. Thus the eventual exploitation of the water-wheel from the third century onwards may well have been stimulated, partly, by the increasingly acute shortage of manpower from which the later Roman empire suffered. Yet again the importance of slavery should not be exaggerated. The ancient slave owner had at least two good reasons to want to reduce his dependence on slave labour if he possibly could, for slaves were quite expensive to feed, and they could be difficult to control.

The contrast between the slow diffusion of the water-wheel and the earlier more rapid exploitation of the Pompeian type of corn-mill is instructive. The Pompeian mill was a rotary mill which had the great advantage over earlier types that it enabled animal power—usually the donkey, but occasionally the horse—to be used. The mill was of fairly simple design (Fig. 23) and could be constructed quite cheaply, and although hand-querns continued to be the usual means of grinding corn in small households, the mill appears to have become established quite rapidly in the western Mediterranean in the second century B.C.[1] In this case, slave labour evidently did not inhibit the exploitation of a mechanical technique. The water-wheel on the other hand was both a good deal more complex, and much more expensive to construct—especially if the water had to be brought to the mill by an aqueduct. The water-wheel needed, as the Pompeian donkey mill did not, a considerable outlay of capital. Although vast sums were spent, both by rulers and by private citizens, on construction works of one kind or another, the ancient world provides very few examples of capital expenditure on a large scale in a manufacturing industry.

Moreover if the donkey mill provides one example of the successful introduction of a new technique, we may put beside it not only the slow diffusion of the water-wheel but also the failure to explore the possibilities of using wind as a source of power, despite the fact that this was familiar to the ancients in one context, that of sailing-ships, and even though the working parts of a water-mill could be adapted relatively easily to be driven by wind. Yet apart from a brief reference to the use of

[1] The history of the development of different types of mills has been described by L. A. Moritz, *Grain-mills and Flour in Classical Antiquity*, Oxford, Clarendon Press, 1958.

wind to work the pump of a water-organ in Hero (*Pneumatics* I 43) there is otherwise no evidence whatsoever that the ancients were alive to the possibilities of harnessing the power of wind.

For reasons that differ in each case, neither steam nor wind was effectively exploited as a source of power in the ancient world, and water only at the end of our period. The main sources of power on which Greek and Roman technology depended were human or animal energy, and this severely limited the scale on which mechanical operations were conducted. Hero's 'automatic theatre', in which puppets go

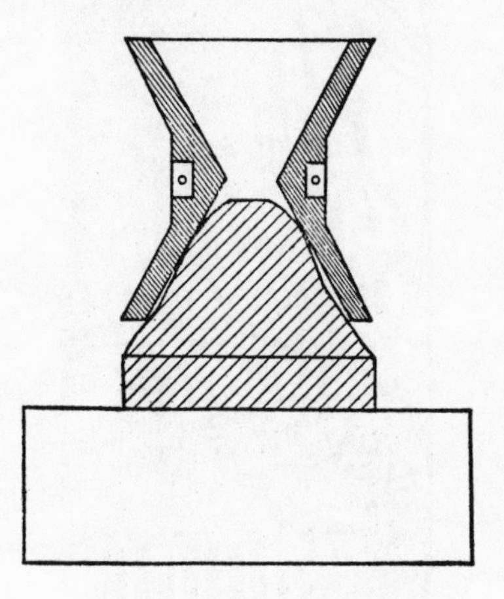

Fig. 23 Pompeian mill.

through the motions of constructing a boat, hammering nails, sawing wood and so on, shows that the idea of automation was not unknown in the ancient world. But whereas the puppet theatre could be driven by harnessing the energy of a heavy weight, there was simply no means of translating this idea into a full scale real life version, and neither Hero nor anyone else attempted to do so. The complex cranes described in Vitruvius, for example, are moved by treadmills worked by slaves (see Fig. 24). The Pompeian corn-mill was normally driven by a donkey. Moreover the effective exploitation of animal energy was in one respect held back in that the ancients never devised a harness that would have enabled the horse to be used

Fig. 24 Part of a relief from a Roman sepulchral monument of about 100 A.D. showing a treadmill being used to work a crane in the building of a monument.

efficiently for traction purposes. The harness they used on the horse was essentially the same as the ox-harness. But while suitable enough for oxen (the main draught animals) this was quite unsuitable for using the horse for a similar purpose, since the breast-band had a tendency, as soon as the horse pulled hard, to slide up the throat and choke the windpipe, thereby drastically reducing the animal's effective power.[1]

[1] The classic study of the harness in antiquity is that of R.J.E.C. Lefebvre des Noëttes (*L'attelage et le cheval de selle à travers les âges*, Paris,

The contempt for the mechanical arts expressed in such writers as Plutarch was only one of several factors that inhibited the development of technology in the ancient world. We should not underestimate the conservative tendencies of ancient technology as a whole, for example in the transmission of technical knowledge. The apprentice was taught to copy the existing methods as exactly as possible. Even in areas such as corn-milling, where developments did take place, long periods when techniques remained unchanged are the rule. The availability of cheap labour was also undoubtedly an important factor. Moreover the comparative success of technology in some fields suggests a fourth consideration to add to those I have mentioned. As the text of Philo referring to artillery shows (above, pp 97 ff), where the power or prestige of rulers was at stake, technology did not lack encouragement. Yet the converse is also to a large extent true: where the advantage of rich individuals was not involved, technology was often neglected. There is a striking contrast between the skills lavished on fine metal-work for instance and the relatively crude methods that continued to be used throughout antiquity in the extraction of metals. In the manufacture of textiles and pottery, too, extraordinarily fine pieces were created, but little attention was paid to solving the problems of mass-production. Wherever they could, in short, the ancients turned their crafts into *arts*: they did not, with few exceptions, attempt to convert them into industries.

Power, honour and status were the most potent motive forces in the ancient world. This is not to say that the profit motive was absent. On the contrary, wealth was clearly actively pursued—however much a Vitruvius might deny that this was his motive for studying the art. Yet as both the source and manifestation of wealth, land was the ideal. Where a man had made a fortune from other sources, commerce or more rarely manufacture, his surpluses were more often devoted to

---

Picard, 1931) some of whose conclusions must, however, be qualified in the light of P. Vigneron's *Le cheval dans l'antiquité gréco-romaine*, 2 vols, Nancy, Faculté des Lettres de l'Université de Nancy, 1968. There is a brief discussion by E. M. Jope in vol 2 of *A History of Technology*, ed. C. Singer and others, Oxford, Clarendon Press, 1956, ch 15.

the purchase of estates than ploughed back (as we should say) into the business. Wealth was, in fact, often treated as a means to an end—a means of entrée to the circle of landed aristocrats —rather than an end in itself.

In a few areas of technology, the engineer could count on the support of powerful individuals or states. The 'architect' had an accepted, though somewhat restricted, place. But outside those areas, the mechanician had to rely on his own resources. He might express the wish to serve practical ends—indeed as we have seen many do so. Yet the idea of systematically exploiting mechanical notions for such ends was generally lacking. His research, like that of other natural scientists, was undertaken as much to satisfy his own desire for knowledge—to understand the phenomena—as for any other motive. Neither curiosity nor ingenuity was lacking: but where no great store was set by material progress in the values of society, the potentiality of applied mechanics to achieve such progress was left largely unexplored.

8

Ptolemy

THUS far in this study we have been concerned primarily with the third and second centuries B.C., an exceptionally fruitful period for almost every branch of Greek scientific thought. The two centuries that followed produced much less important original work. In the second century A.D., however, two major figures, who represent in many ways the culmination of ancient science, provide an opportunity to take stock of its achievements in the fields in which they worked. These are Ptolemy and Galen—the one primarily an astronomer, the other mainly a biologist—the very success of whose work is largely responsible for the difficulties we have encountered in reconstructing that of some earlier scientists whose contributions they eclipsed.

Ptolemy's *Mathematical Composition*, which has generally been referred to, since the Middle Ages, by its Arabic name, the *Almagest*,[1] is the most comprehensive astronomical treatise that has come down to us from antiquity. Whereas the astronomical work of Eudoxus, Aristarchus, Apollonius and Hipparchus is known to us only from fragments or, at best, from minor treatises, the whole of Ptolemy's masterpiece has survived, as well as other works of his, including several on astronomical subjects (for example, *The Hypotheses of the Planets*, and the astrological work known as the *Tetrabiblos*), the *Geography*, and treatises on music and optics.

About the author himself we know very little. The references in the *Almagest* to the astronomical observations that he carried out 'in the parallel of Alexandria' establish that he lived in Egypt, almost certainly at Alexandria itself, and they also enable us to fix the approximate dates of his main astronomical work. The earliest of his own observations is 127 A.D.,[2] and they continued at least until 141 A.D.

[1] This is a corruption formed by prefixing the article *Al* to the Greek superlative adjective *megistē*, 'the greatest'.

[2] Ptolemy refers to this as year eleven of Hadrian: alternatively it

Ptolemy knows his predecessors well and uses them extensively, frequently expressing his admiration for Hipparchus in particular. In the opening chapter of the *Almagest* he writes:

> We shall try to set down as concisely as possible whatever we consider to have come to light up to the present. . . . So as not to make the treatise too long we shall only report what was exactly investigated by the ancients, but we shall perfect as far as we can what was not comprehended completely or not as well as possible,

and he describes himself as 'making such an contribution [to knowledge] as the time that has elapsed between them and the present might well be able to make possible'. But Ptolemy's modesty here should not mislead us into supposing that he was a mere eclectic. Apart from his achievements in observational astronomy—where his star-catalogue, for example, though largely dependent on the work of Hipparchus, was more comprehensive than any earlier description of the heavenly bodies[1]—he was responsible for several innovations in the theory of the movements of the moon and planets.

Ptolemy describes the aims of astronomy at the outset of his work. He mentions the Aristotelian distinction between theoretical and practical studies and subdivides the theoretical into three, theology—the study of god (conceived as invisible and unchanging)—'physics', here the study of the world of change in the sublunary sphere, and 'mathematics', including, especially, theoretical astronomy. But both theology and physics are, he says, matters of conjecture rather than of scientific understanding, theology because of its utter obscurity, and physics because of the instability of what it deals with. 'Mathematics' alone yields unshakeable knowledge, proceeding as it does by means of indisputable arithmetical and geometrical demonstrations. The purpose of the study is not

---

is year 874 of Nabonassar, the first year of whose reign (747 B.C.) is the base line of Ptolemy's system of calculating dates.

[1] 1,028 stars are included, and according to the calculations made for the year 100 A.D. by C. H. F. Peters and E. B. Knobel in *Ptolemy's Catalogue of Stars*, Carnegie Institution of Washington, 1915, the mean error of longitude is about 51′ and of latitude about 26′.

merely to obtain knowledge, but also to appreciate the beauty and order of the heavenly bodies, and indeed Ptolemy claims that astronomy improves men's characters:

> of all studies this one especially would prepare men to be perceptive of nobility both of action and of character: when the sameness, good order, proportion and freedom from arrogance of divine things are being contemplated, this study makes those who follow it lovers of this divine beauty and instils, and as it were makes natural, the same condition in their soul. (Book I chapter 1)

Ptolemy proceeds to formulate and justify the fundamental physical theses on which his astronomical system is based, such as the sphericity of the heavens, the sphericity of the earth and the doctrine that the earth is at rest in the centre of the universe. Once again the influence of Aristotelian doctrines is evident. In putting forward the first two theses in book I chs 3 and 4 Ptolemy uses arguments of very different types. Thus to prove the sphericity of the heavens, he refers not only to the evidence of the observed circular motion of the circumpolar stars, but also to such physical considerations as that the heavens are composed of *aithēr*: this is the most homogeneous element, and since the surfaces of homogeneous bodies will themselves be homogeneous, and the most homogeneous solid figure is the sphere, we may suppose that *aithēr* is spherical in form.

Chapter 5 sets out to establish that the earth is at the centre of the universe by refuting all other possible positions. His most telling argument is that if the earth were not at the centre, the plane of the horizon would not bisect the stellar sphere—whereas in fact both the celestial equator and the ecliptic are bisected by the horizon. This point was, indeed, to present a difficulty to Copernicus when he revived the heliocentric theory, though he was able to meet it by appealing to considerations already anticipated by Ptolemy himself. First, as Ptolemy knew full well, even on the geocentric theory, the position of the observer, on the surface of the earth, does not coincide exactly with the centre of the universe. Secondly, Copernicus saw that although the heavenly sphere would not be bisected by the horizon if the earth were at a considerable distance from the centre of that sphere—considerable, that is, in relation to the size of the sphere itself—the correct conclusion to draw is not that the centre of the earth coincides with that of the universe, but rather that the heavenly sphere is immensely great in comparison with the earth, a thesis that Ptolemy himself upheld in I ch 6, when he argued that the earth has the ratio of a point to the sphere of the heavens.

Ptolemy then turns, in chapter 7, to the question of whether the earth moves or is at rest, and this chapter provides our chief source for the arguments that were used in antiquity, some of which have already been mentioned on pp 58 ff, to reject the idea that the earth is subject to any movement whatsoever. First, he repeats Aristotle's main argument that heavy bodies must always and everywhere move in the same direction. They move along straight lines drawn to the centre of the universe—which coincides, as Ptolemy had just argued, with the centre of the earth. 'So it seems to me, at least,' he writes,

> superfluous to seek the causes of the motion towards the centre, when once the fact that the earth occupies the middle position in the universe and that all weights move towards it, is made so clear from the appearances themselves.

The earth is of huge size in comparison with the things that fall on it, nor is it affected by the impact of the weights falling on it. Moreover if the earth had any natural motion, its speed would, according to the principles of Aristotelian dynamics, be proportional to its weight, and so it would long ago have left behind all the objects near it:

> if [the earth] too had had a single common motion, the same as that of the other weights, it would clearly have got ahead of everything as it fell because of its vastly greater size; and the animals and all separate weights would have been left hanging in the air, and it would very quickly have fallen completely out of the universe itself. But one has only to think of this sort of suggestion for it to appear utterly absurd.

The hypothesis of the axial rotation of the earth is then considered and again rejected largely on the grounds that the speed of the movement would have to be so great. Yet he remarks that 'so far as the phenomena relating to the stars are concerned, perhaps nothing might prevent things from being in accordance with the simpler [form of this] theory'. But he goes on:

> yet to judge from what happens on earth and in the air about us, such an idea would be seen to be utterly absurd... [Advocates of this view] would have to admit that the rotation of the earth would be the most violent of all the simple movements about it... No cloud would ever be shown travelling towards the east, nor would anything else that flies or is thrown, since the earth would always be anticipating them and forestalling their motion towards the east, so that everything else would seem to be left behind and recede towards the west.

He recognizes that those who maintained the axial rotation of the earth had a line of defence in the suggestion that the air too is carried round with the earth. But to this he replies that solid bodies moving through the air would still appear to be left behind. And to the suggestion that the solid bodies themselves might move round with the air and the earth, being attached as it were to the air, Ptolemy comments that this would make any change in the relative positions of solid bodies in the air impossible. They would never 'appear either to move forwards or to be left behind. ... They would never wander nor change their position either in flying or in being thrown, although we so clearly see all these things being brought about, just as if no slowness or swiftness whatever accrued to them as the result of the earth not being at rest.'

On this important problem, then, Ptolemy draws heavily on Aristotelian physical doctrines, but he believes these to be supported by the evidence of observation. He refers repeatedly to the 'phenomena' or the 'appearances', the observed behaviour of falling bodies, or missiles, or other objects travelling through the air. To detect the effects of the earth's rotation would, indeed, have required much more accurate instruments than were available to Ptolemy. Crude observation of the movement of objects near the earth provided no grounds to doubt that the earth is at rest, and a wealth of data that seemed to point to that conclusion. Ptolemy does not consider the heliocentric hypothesis as such, but in his view that was open not only to the physical objections he mentioned in refuting the idea that the earth moves in space, but also to the astronomical arguments he brought in chapter 5 against any theory that removed the earth from the centre of the heavenly sphere.

Finally among the preliminaries in book I he sets out the Table of Chords and the basic trigonometrical propositions that he will use in the rest of the *Almagest*.[1] The Table of Chords serves the same purpose as a table of sines or cosines. Where we refer to sines and cosines, the Greeks referred to the

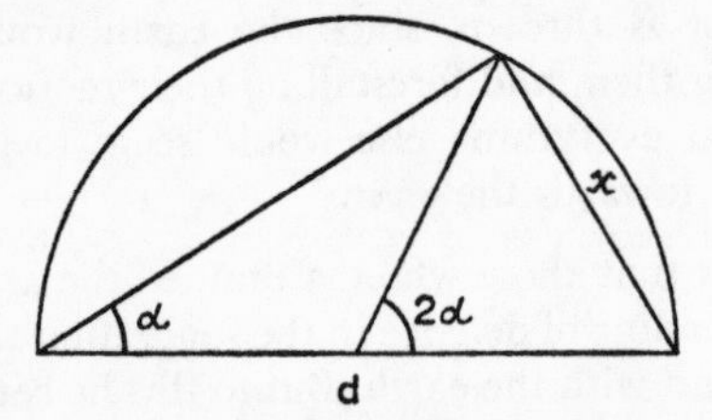

Fig. 25 Chords in a circle.

chords subtended by arcs of a circle: the chords are expressed as so many $\frac{1}{120}$th parts of the diameter of the circle, but as sexagesimal fractions are used, this is numerically equivalent to division by two: it follows that our sin α is the same as the Greek $\frac{1}{2}$ chord 2α.[2] The table in I ch 11, which was worked out with the help of what is known as 'Ptolemy's theorem', which enables chord $(\alpha-\beta)$ to be found, given chord α and chord β, sets out values for chords for every half a degree up to 180°.

[1] The most important extant text for Greek trigonometry before Ptolemy is the *Sphaerica* of Menelaus of Alexandria (end of the first century A.D.) which has been preserved in an Arabic version. But Ptolemy's chief debt—here as elsewhere—is to Hipparchus, who appears to have been the first person to construct a Table of Chords and who is reported to have written a treatise in twelve books *On Chords in a Circle*.

[2] In Fig. 25, $d \sin \alpha = x =$ (in Greek terminology) $\frac{d}{2}$ chord 2α.

Like his predecessors, Ptolemy conceives the main task of the astronomer to be to 'save the appearances' by explaining the apparently irregular movements of the heavenly bodies in terms of combinations of uniform circular motions, and in general he takes as his starting-point the Apollonian–Hipparchan model of epicycles and eccentrics. In his theory of the sun, for instance, he evidently follows Hipparchus closely. In book III Ptolemy shows that the irregularities of the sun's movement, seen in the inequalities of the seasons, may be explained either on the hypothesis that the sun moves on a circle eccentric to the earth, or on the hypothesis that it moves on an epicycle. He sets out the proof (that probably goes back to Apollonius, see above, p 63) that these two hypotheses are equivalent, and he expressly prefers the eccentric model on the grounds of its greater simplicity, since it presupposes one movement rather than two.

In his accounts of the moon and the planets, however, Ptolemy introduces several important modifications into astronomical theory. In book V, the second of three books devoted to the moon, he first describes how he determined its position by using the 'astrolabe'—that is the armillary astrolabe, a sighting instrument not to be confused with the plane astrolabe described, for example, by Philoponus. The armillary astrolabe is the most important and complex of the instruments that Ptolemy used (see Fig. 26) and it had the great advantage that, once the instrument was set on a known fixed point (the sun, the moon, or a fixed star), it enabled the ecliptic coordinates (latitude and longitude) of a heavenly body to be determined directly, rather than by complicated calculations from observations of its position in relation to the zenith and the horizon. While some form of this instrument very probably antedates Ptolemy, he tells us that he constructed his own example and he claims that his use of it, to observe the moon, brought to light certain discrepancies between its actual positions and those predicted on the current theory, which itself had been based largely on data derived from observations of eclipses.

> In general, [he writes in V ch 2] observing in this way . . ., we found that the distances of the moon in respect of the sun

> sometimes agreed with the calculations made according to the hypothesis we have expounded, but sometimes differed and disagreed with them, at times by a little, at times by a great deal.

'But,' he continues, 'carrying out a progressively more complete and more meticulous examination', it was found that at conjunctions and at full moons there was little or no discrepancy. But when the moon was in the first or third quarter, and at the same time midway between the apogee and the perigee (that is, the furthest and nearest points from the earth) on the

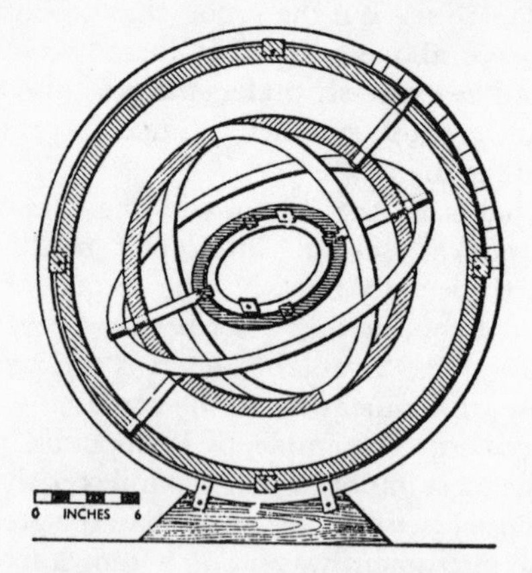

Fig. 26 The armillary astrolabe.

epicycle, the discrepancies between what was observed and what was predicted were appreciable (see below Fig. 27).

The way in which Ptolemy dealt with this problem is strictly in line with the way in which the epicycle model itself dealt with variations from the mean, or true circular, path of a heavenly body. Assuming that the moon moves on an epicycle which is carried round on a deferent circle, he suggested that the centre of this deferent circle is not (as had usually been held) the earth itself, but a point which itself revolves in a small circle with the earth as centre. In the figure (27), the centre of the epicycle (C) is at a fixed distance from the centre of the deferent, the movable point F, and this point rotates about the earth (E) in a sense contrary to that of the motion of the centre of the epicycle round F. The new model agrees with the old when the moon is at conjunction or in opposition (when the sun, moon and earth are in a straight line). But the new model could be used to account for the observed discrepancies when the moon is in the first and third quarters.

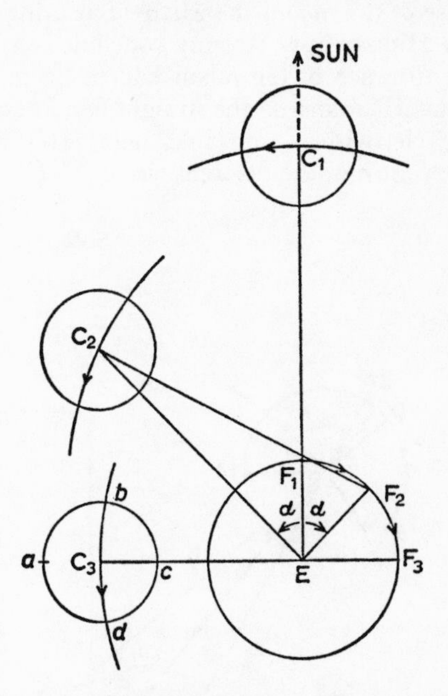

Fig. 27 Ptolemy's model to explain the moon's second anomaly. The moon's epicycle, centre C, moves round a centre (F) which itself describes a circle round the earth (E). F moves round E with the same angular velocity as, but in the contrary sense to, the movement of C round E.

At position (1) the model is equivalent to a simple epicycle model. At position (3) when the moon is at apogee or perigee on the epicycle (that is, at *a* or at *c*), the model is again equivalent to a simple epicycle model so far as the moon's angular distance from the sun is concerned. But at position (3) when the moon is at *b* or *d*, midway between apogee and perigee on the epicycle, the effect of the new model is to increase the apparent diameter of the epicycle.

But having introduced this first modification into lunar theory, Ptolemy went on to suggest a second—the doctrine of 'direction'. This concerns the angular distance of the moon on its epicycle. The normal rule, in the epicycle theory, was that the angular distance of the heavenly body on its epicycle was measured from the diameter lying on the line joining the centre of the epicycle with its centre of regular rotation (that is, in the case of the moon, the earth). But using observations recorded by Hipparchus, Ptolemy concluded in V ch 5, that the angular distance of the moon had to be measured from another point (H) found on the straight line joining the centre of the epicycle with a movable point (N) diametrically opposite the centre of the deferent circle (F) (Fig. 28).

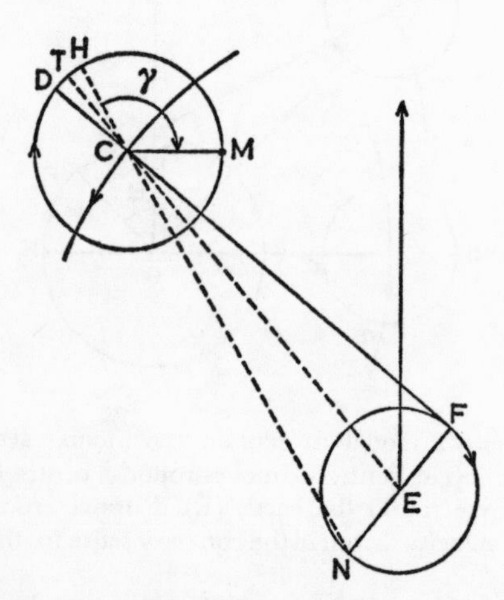

Fig. 28 The doctrine of 'direction' (after Neugebauer). The angular distance (γ) of the moon (M) on its epicycle is measured not from D (on the line joining the centres C and F) nor from the 'true' apogee (T) but from a variable point (H—the 'mean' apogee) found on a straight line joining the centre of the epicycle (C) with a point (N) situated diametrically opposite the centre of the deferent circle (F).

For the planets, where, as we have seen (p 67), Hipparchus had not been able to arrive at a satisfactory general theory, Ptolemy introduced similar modifications to the original epicycle model. Very briefly, his theory of the movement of all the planets except Mercury is broadly the same. The planet (P in Fig. 29) is imagined as moving on an epicycle, the centre of which moves round a deferent circle. The centre of this

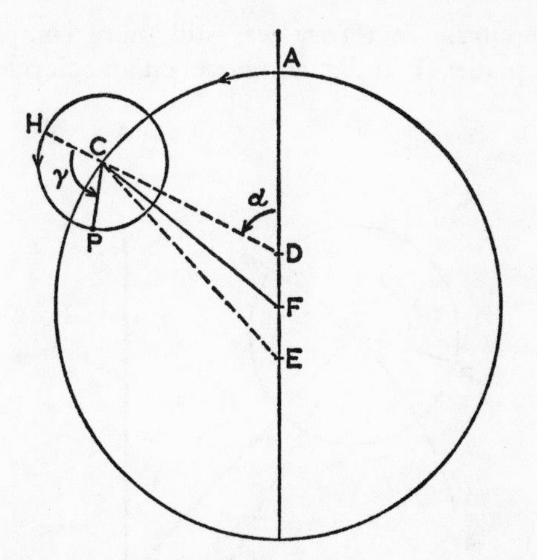

Fig. 29 Ptolemy's model for the planets other than Mercury (after Neugebauer). The planet P moves on the circumference of an epicycle centre C. C moves round an eccentric circle, centre F, but the movement is uniform not with respect to F, but with respect to D ('the equant')—the angle α increases uniformly. The equant is a point on the line from the earth (E) through the centre of the eccentric circle (F) and such that EF = FD.

circle (F) is fixed, but it does not coincide with the earth: that is, the deferent circle is an eccentric one. Moreover the centre of the epicycle moves uniformly not with respect to its own centre (F), nor with respect to the earth (E), but with respect to a point D (later called the equant) on the straight line EFA and such that DF = FE. The line DH turns with uniform angular velocity round D, and the planet's movement on its epicycle is measured from the same line. The longitude of the planet depends, therefore, on these two variables (α and γ in the diagram). The rotation of the epicycle is in the same sense as that of the deferent circle, which produces the phenomenon of retrogradation (see above, figure 9): the planet has its greatest apparent easterly speed near the apogee (when it is furthest away from the earth on its epicycle) and retrogradation occurs when the planet is near perigee on its epicycle.

For Mercury, the theory was still more complex. Once again, the planet (P in Fig. 30) moves on an epicycle, but now

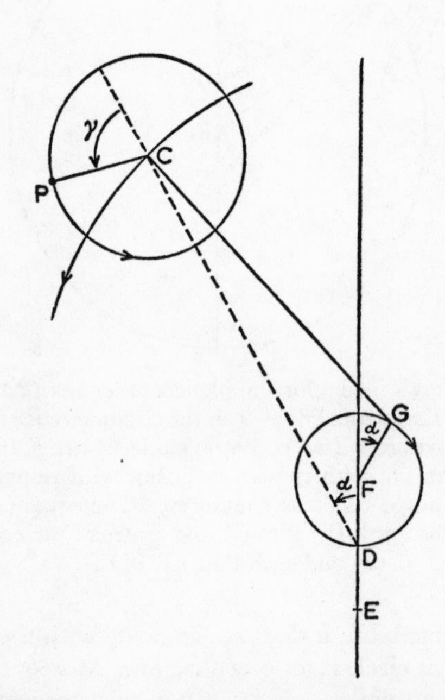

Fig. 30 Ptolemy's model for Mercury (after Neugebauer). The planet (P) moves on an epicycle, whose centre (C) moves round an eccentric circle with movable centre (G). G moves round the circumference of a circle, centre F, at the same speed as, but in the opposite sense to, the motion of C on the deferent circle measured with respect to D.

the epicycle's centre moves on a circle whose centre (G) is not only eccentric to the earth (E), but itself in motion round a circle centre F, found on the straight line EDF and such that ED = DF. G revolves round F at the same speed as, but in a contrary sense to, the revolution of the epicycle's centre on the deferent circle measured with respect to the equant D. In the diagram the angle GFA is always equal to the angle FDC. It follows that in one revolution of the deferent circle, there is one apogee (when AGFDE is a straight line) but two perigees (when angle GFA is about 120°).

The system as a whole is geocentric: yet for each of the planets one of the two main movements governing its position is linked with the sun. In the case of the three outer planets (Saturn, Jupiter, Mars) the line from the planet to the centre of its epicycle remains parallel to the line from the earth to the sun, while for the two inner planets (Venus, Mercury) the centre of the epicycle is thought of as on that line (see Fig. 31). From the point of view of the heliocentric theory, then, we may say that for the three outer planets the epicycle represents the annual motion of the earth round the sun, and the deferent corresponds to the motion of the planet itself round the sun. For Venus and Mercury the reverse applies: the epicycle corresponds to the movement of the planet round the sun and the deferent to the annual revolution of the earth.

Having set out the main elements of his theory in book IX ch 5 and 6, Ptolemy proceeds in the rest of that book and in the next four to give a systematic account of each planet in turn. Thus for each planet he determines, by calculations based on observations that he cites, (1) the size of the planet's epicycle, (2) the planet's eccentricity (both these are expressed as proportions of the radius of the deferent circle) and (3) tables from which the longitudinal position of the planet can be computed,[1] as well as (4) the magnitudes and durations of the retrogradations of each planet.

Ptolemy's theory of the moon and the planets offers a

[1] Having set out, in IX 3, the periodic returns of each of the planets (that is, the periods of revolution of the deferent circle and of the epicycle) he explains in XI 10 and 12, how to compute the longitudes of the planets from the tables he gives in XI 11. The latitudinal movement of the planets is dealt with separately in book XIII.

complete determination of their movements. Yet the theory as a whole was open to two principal types of criticisms. First, it could be objected that when the motion of a circle was held to be uniform with respect to some other point than its own centre, as in the doctrine of the 'equant', this was tantamount to a breach of the fundamental principle according to which the movements of the heavenly bodies should be explained in terms of combinations of regular circular motions. Indeed Copernicus was to criticize Ptolemy's lunar theory both on

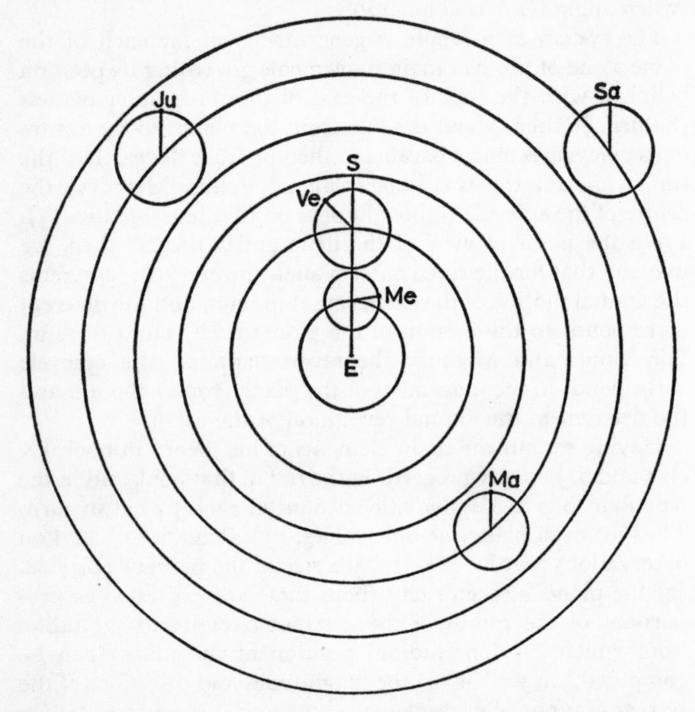

Fig. 31 Simplified representation (ignoring eccentricity and equants, and not to scale) to show relationship between planets and sun in Ptolemy's system. The centres of the epicycles of Venus (Ve) and Mercury (Me) are on the line joining the Earth (E) to the Sun (S). For the outer planets, Saturn (Sa), Jupiter (Ju) and Mars (Ma), the line joining the planet to the centre of its epicycle remains parallel to ES.

that score and on the grounds that the doctrine of 'direction' is irregular. Thus in book IV ch 2 of his *De Revolutionibus Orbium Coelestium* (1543) Copernicus remarks that the movement of the epicycle on the eccentric circle is not regular and asks:

> If that is the case, what reply should we give to the axiom that the movement of the heavenly bodies is regular except that it may seem irregular with respect to appearances—if the apparent regular movement of the epicycle is in fact (*reipsa*) irregular and it takes place quite contrary to the principle set up and assumed?

And again on 'direction' he says: 'As also the moon traverses its own epicycle irregularly, if we should now try to confirm the irregularity of the appearances by means of irregular movements, we may remark what sort of argument that would be.'

Secondly, both the lunar and the planetary theories ignore certain data. By far the most striking example of this concerns the values Ptolemy assigned to the magnitudes of the circles that determine the course of the moon. Its course was, as we have seen, the resultant of the combined movements of three circles. But the values that Ptolemy gave to the radii of these circles had one consequence that conflicted directly and blatantly with what is observed. From his account, it followed that the moon's distance from the earth varies by as much as 34 to 65, or nearly 1 to 2. Since, for small angles, the tangents are nearly proportional to the angles, this means that the apparent diameter of the moon at perigee should be almost twice its apparent diameter at apogee. Yet this is obviously not the case. Moreover Ptolemy knew this very well: he gave a tolerably accurate estimate of the moon's diameter at maximum distance from the earth in V 14, namely 31′ 20″, and his figure for the diameter at minimum distance is quoted by Pappus as 35′ 20″. Yet in setting out his doctrine of the moon's movements, he passes over this in silence.

The reason for this is, however, fairly clear and it throws some light on Ptolemy's aims and assumptions in the bulk of the *Almagest*. His immediate purpose was to construct a model that can account for the movements of the moon in longitude and latitude and enable its positions to be predicted. At this point he was not interested in attempting to give a *physical* model to explain the actual causes governing the moon's motion. For the former purpose his model is sufficiently exact, even though if treated as a physical account, it is at one point in glaring contradiction with the facts.

Ptolemy ignores this difficulty, even though elsewhere physical considerations are a determining factor in his astronomy. We have seen that he rejected the doctrine of the earth's rotation for what are largely physical reasons, even though 'so far as the phenomena relating to the stars are concerned, perhaps nothing might prevent things from being in accordance with the simpler [form of this] theory' (see above p 117). The astronomical system in the *Almagest* is firmly set in the framework of certain physical assumptions, mostly Aristotelian doctrines, and he is not prepared to abandon these, despite the economy that would have resulted in his calculations. Again the work on *The Hypotheses of the Planets* shows that he hoped for a physical account. Here the circles that govern the movements of the heavenly bodies are conceived as strips of spheres, though in describing the causes of their motion he rejects the Aristotelian picture of interacting spheres and falls back on vitalist notions: 'we have to suppose that among the celestial bodies each planet possesses for itself a vital force and moves itself and imparts motion to the bodies united with it by nature'.[1] Yet although his ultimate goal is, no doubt, an account that is both mathematically exact and physically true, in the body of the *Almagest* he concentrates on the former aim, namely to provide a geometrical model from which the motions of the sun, moon and planets can be computed.

The role that the notion of simplicity plays in his thinking comes out again when he returns, in book XIII ch 2, to attempt a solution to the problems posed by the fact that the orbits of the planets deviate from the ecliptic:

> Let no one think such hypotheses troublesome, considering the inadequacy of the devices open to us. For it is not fitting to compare human things with divine ones, nor to

[1] Book II ch 7. This book has been preserved only in an Arabic translation from the Greek original. The English translation in my text is that of Sambursky, *The Physical World of Late Antiquity*, London, Routledge and Kegan Paul, 1962, p 144.

> derive arguments concerning such great things from examples that are so unlike them. . . But we should try to fit the simpler hypotheses, as far as possible, to the movements in the heavens, and if this does not succeed, then any possible hypotheses. For once each of the appearances is saved as a consequence of the hypotheses, why should it still seem strange that such complications can occur in the movements of the heavenly bodies? . . . Rather it is not fitting to judge the very simplicity of heavenly things from those that seem to be simple with us, when not even with us is the same object equally simply to everyone alike . . . Rather [we should judge their simplicity] from the unchangeableness of the natural substances in heaven itself and of their movements. For thus all would appear simple, and more so than those things that seem so with us, since it is unthinkable that there is any labour or difficulty in their revolutions.

We see from this that the notion of simplicity operates in two quite distinct ways. On the one hand Ptolemy appeals to it as a *criterion* for preferring one model to another. On the other, recognizing the complexities of the appearances and of the devices he used to 'save' them, he argues that the notion of simplicity that we derive from experience of the sublunary world is inadequate and inappropriate. The heavenly bodies are unchanging, and it is *axiomatic* that, however they appear to us, their movements are simple.

Yet once this has been said, it must be emphasized that it is only as a last resort that Ptolemy appeals to this Aristotelian distinction between the divine and the sublunary world. Despite its shortcomings—and he recognizes both that there are problems that remain unsolved and that some of his devices are arbitrary—the *Almagest* is an extraordinary achievement, for the rigour of its mathematical arguments, for the range of data encompassed and the comprehensiveness of the results proposed. Although the abandonment of geocentricity led eventually to the possibility of a new synthesis, there is an essential continuity in astronomical theory from Ptolemy through the Arab astronomers to Copernicus. As Neugebauer has put it, comparing the *Almagest* with the *Opus astronomicum* of the early tenth century Arab astronomer Al-Battani and Copernicus' *De Revolutionibus*, 'there is no better way to convince oneself of the inner coherence of ancient and mediaeval astronomy than to place [these books] side by side.... Chapter by chapter, theorem by theorem, table by table, these works run parallel. With Tycho Brahe and Kepler the spell of tradition was broken.'[1]

The most extensive of Ptolemy's other astronomical works, apart from the *Almagest*, is the astrological treatise, in four books, known as the *Tetrabiblos*. Many distinguished scholars have, in the past, flatly refused to believe that this could have been written by the same man who composed the *Almagest*. Yet we have no reason to doubt its authenticity. The language and style of the two works, and the framework of astronomical doctrine in them, are the same, and there are clear references to the *Almagest* both in the introduction and in the body of the *Tetrabiblos*. Nor should we be in any way surprised that Ptolemy should be interested in astrology, when we reflect that the same is true of so many other astronomers in antiquity and in the Renaissance, including Copernicus, Tycho Brahe, Kepler and Newton. What is more interesting is to consider what Ptolemy has to say concerning the relation between the two types of investigation.

The term *astrologiā* can cover both what we should call astronomy and what we should describe as astrology. The term *mathēmatikos* too was used by Ptolemy, as by other ancient writers, for the astrologer as well as the astronomer. But that is not to say that he failed to draw any distinction between, on the one hand, the analysis of the courses of the heavenly bodies and the attempt to predict their movements, and, on the other, the attempt to determine their supposed influence on the sublunary world in general and on individual men in particular. On the contrary, that distinction is clearly and definitely drawn in the opening chapters of book I of the *Tetrabiblos*. Moreover Ptolemy is sceptical about many features of his predecessors' practice of astrological divination. He rejects some parts of their 'art' outright, as, for example, when he says in book III: 'we shall dismiss the superfluous nonsense of

[1] O. Neugebauer, *The Exact Sciences in Antiquity* (second edition), Providence, R. I., Brown University Press, 1957, pp 205 f.

the many, that lacks any plausibility, in favour of the primary natural causes.' Again he makes a more fundamental point when he remarks that predictions concerning the sublunary world are bound to be uncertain and conjectural, since the nature of the subject-matter they are dealing with does not permit certainty. Yet although he shows a certain caution in these respects, he does not doubt the possibility of astrological divination in general. Along with other, less plausible, examples, he cites such *prima facie* evidence for the influence of the heavenly bodies on the earth as the effect of the sun on the seasons and that of the moon on the tides. Predictions are possible, he believes, even if the stars are difficult to read. Although he confines himself to generalities—discussing, for instance, which heavenly bodies are beneficent, which maleficent, rather than himself attempting predictions concerning individual men—these generalities extend to four books and are intended to provide the framework within which divination concerning individuals can be practised.

Once again we must recall that the question of which studies can, and which cannot, lay claim to being genuine branches of knowledge was itself, in the ancient, as in the modern, world, a controversial one. Ptolemy believes that there is no reason, in principle, why general predictions cannot be made about events on earth on the basis of observations of the movements of the heavenly bodies. Precise and certain predictions are ruled out. But the reason for that lies not in the nature of the evidence, so much as in the nature of sublunary matter itself. Like Plato, he holds that in studies concerning physical objects no certainty is possible, and for the same reason he is no fatalist, since he believes that 'lesser' causes yield to 'greater' ones. As he says of astrology in book I ch 2, 'every study that deals with the quality of matter is conjectural'. It is, however, a proper subject of study and he justifies it on the grounds that the knowledge it aims to provide enables a man to face the future with calm and steadiness.

Apart from the study of the heavenly bodies, Ptolemy was interested in a wide range of subjects, including geography (which he divided into a descriptive study—'chorography'—and a theoretical, mathematical branch—geography proper—concerned particularly with the problems of projection), acoustics and musical theory, and optics. His work in this last subject is especially interesting for the light it throws on his methods. Even though our source—a Latin translation of a twelfth century Arabic version of his treatise—is at two removes from Ptolemy's text itself, it enables us to assess certain features of his investigations.

At the beginning of book III, for instance, Ptolemy sets out three elementary principles of optics. These are (1) that objects that are seen in mirrors appear in the direction of the visual

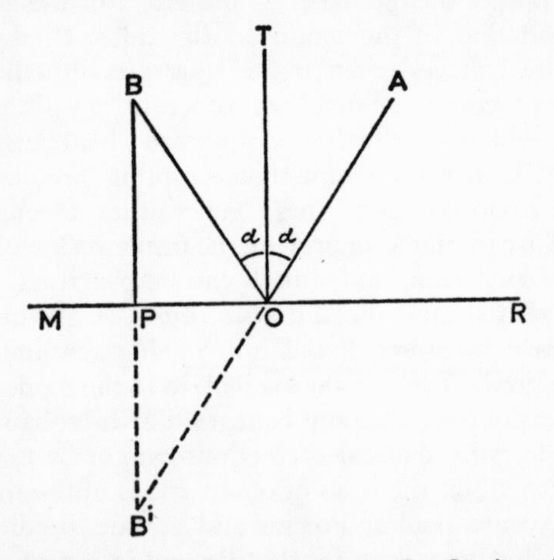

Fig. 32 Ptolemy's elementary principles of reflection. MR is the mirror, A the eye, B the object, B^1 the image, O the point at which the visual ray strikes the mirror and TO and BP perpendiculars to the mirror.

ray which falls on them when reflected in the mirror (the ray being considered as proceeding *from* the eye); (2) that images in mirrors are seen on the perpendicular drawn from the object to the surface of the mirror and produced; and (3) that the position of the reflected ray, from the eye to the mirror and from the mirror of the object, is such that each of its two parts contains the point of reflection and makes equal angles with the perpendicular to the mirror at that point. In the Fig. (32), these three principles are (1) B^i lies on AO produced, (2) B^i lies on BP produced, and (3) angle TOA = angle TOB.

The truth of these principles had, no doubt, been known long before Ptolemy: his contribution is to confirm them by experiment. Thus to establish his first principle he suggests an experiment in which the points on the surface of a mirror through which the image appears are marked, and then covered: 'then the image of the object will certainly no longer be visible. But then when we uncover the points one by one and look at the uncovered points, both the points and the image of the object will be seen together on the straight line drawn to the summit of the visual ray [i.e. the eye].' Further tests follow, including one that confirms, by measurement, that the angle of incidence is equal to the angle of reflection with plane, convex and concave mirrors.

Similar methods are used in his discussion of refraction in book V. He first points out that as in reflection, so too in refraction, the image is seen at the intersection of the line of the visual ray and the perpendicular drawn from the object to the reflecting or refracting surface, and he refers to a test that goes back at least to Archimedes, in which a coin is put into an opaque vessel in such a position that it is just hidden by the lip of the vessel, but comes into view when water is poured into the vessel. But not content merely to state certain general principles of refraction, he undertakes detailed investigations to measure the amount of refraction that takes place at different angles of incidence, and in different media. Thus in section 7 ff he first describes his apparatus. To measure the angles he uses a circular disk, each quadrant of which is divided into ninety parts, like a protractor. This disk is set up in a bowl of clean water so that the water just covers the bottom half of the circle. A coloured marker is placed at a given point (e.g. 10°) along the circumference of the semicircle which is above the water, and the marker, the centre of the disk and the eye are aligned. A small, thin rod is then moved along the circumference of the opposite quadrant (under the water) until the extremity of the rod appears in line with the coloured marker and the centre of the disk (see Fig. 33). This enables both the angle of incidence[1] and the angle of refraction to be determined, and Ptolemy remarks that the former is always

[1] Ptolemy considers the ray from the eye (not that from the object) the incident ray.

greater than the latter, and that as the former increases the amount of refraction also increases. He sets out his results in detail: when the angle of incidence is 10°, the angle of refraction will be about 8°; when the angle of incidence is 20°, the angle of refraction will be $15\frac{1}{2}$°, and so on for angles of incidence up to 80°. Concluding the passage with the remark that 'this is the method by which we have discovered the amount of refraction in water', Ptolemy adds: 'we have found no perceptible difference between waters of different densities and

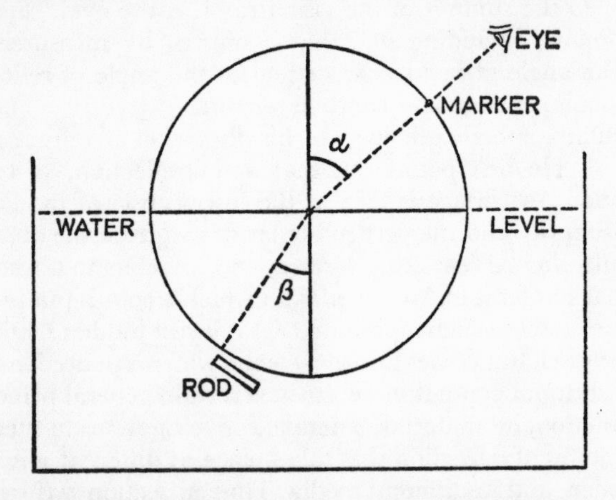

Fig. 33 To illustrate Ptolemy's investigation of refraction between air and water. Each quadrant of the disk is graduated into 90°. α is the angle of incidence, β the angle of refraction.

rareties', and he proceeds to a similar investigation of the refraction of other media, namely from air to glass and from water to glass.

Ptolemy's studies of reflection and refraction involve deliberate and systematic experimentation. Three points stand out. First, in this case experiments are comparatively easy to conduct. In this respect, elementary optics differs markedly from chemistry or dynamics, for example. Secondly, the experiments were evidently repeated and their conditions varied. Thus it appears from the remark just quoted that in his investigation of the refraction of water he experimented with waters of different densities. Thirdly, an analysis of his results shows that they have been adjusted. They are given correct to within half a degree: what is more important, although he states no general law of refraction, it is clear that he conceived the relation between the angle of refraction (r) and the angle of incidence (i) to be of the form $r = ai - bi^2$ (where a and b are constants that depend on the specific media). In stating his results, Ptolemy has evidently corrected them to tally with his general law. He has allowed, one might say, for 'experimental error', or rather he trusts his general, mathematical law more than particular observations. As so often in ancient science, observation is subordinated to theory. In astronomy, as has often been remarked, the Greeks were, on the whole, quite right to place greater confidence in their excellent mathematical methods than in observations that were, given the instruments they used, necessarily inexact. But we find a similar tendency also in optics. At the same time Ptolemy's meticulous investigations of reflection and refraction prove that he by no means devalued the use of observation. On the contrary, his preference for theory over observation is compatible with, and indeed accompanied by, a keen awareness of the importance of the latter. His detailed account of his optical investigations shows that he undertook extended observations to assemble the data—even if these data were then corrected, silently, in the light of the theories that he believed they established.

9
Galen

THE second figure who dominates science in the second century A.D. is Galen of Pergamum, about whose life we know much more than is usual for ancient scientists. Many of his works contain anecdotes that throw light on his career and personality, and three in particular provide information about his life, namely *On Prognosis*, and the two short books where he lists his own treatises and gives guidance about the order in which they should be read, *On his own books* and *On the order of his own books*.

He was born in 129 A.D. at Pergamum. His father, Nicon, was an architect who himself educated his son in mathematics, grammar, logic and philosophy, but who, when Galen was 16, was persuaded by a dream to arrange for his son to be taught medicine also. Galen travelled to Smyrna, Corinth, and Alexandria in connection with his studies and in 157 returned to Pergamum and took up appointment as surgeon to the gladiators, a post he held for four or five years. Then followed a visit to Rome where he set about establishing a reputation in what was, on his own account, not only a highly competitive, but also a corrupt, profession. This visit lasted three years, when Galen, disgusted at the envy and backbiting of his colleagues, and hearing that the civil strife at Pergamum had come to an end, returned to Asia Minor. However not long after he arrived back at Pergamum, letters arrived summoning him to attend the emperors, Marcus Aurelius and Lucius Verus, who were at that time (168) planning a campaign against the Germans. Galen met the army at Aquileia, but when plague attacked the troops, the emperors and their immediate entourage left for Rome and Galen was left to spend a miserable winter with the disease-stricken army. Asked by Marcus Aurelius to accompany his German expedition, Galen pleaded successfully to be allowed to remain at Rome as physician to the emperor's son Commodus. In 192 Galen had the misfortune to have many of his treatises destroyed when the temple of Peace, in which they were deposited, was burnt. By now, however, he was well established and he continued to enjoy imperial favour for the rest of his career. The date of his death is not certain: it is commonly taken to be 199 or 200, but the testimony on which this relies is not strong, and some of our evidence suggests that he lived some ten years or more after the turn of the century. What is clear is that he had a full and successful life devoted to medical practice and authorship.

Although Galen's chief work was in biology and medicine, he was known also as a philosopher and philologist. Training in philosophy is, in his view, not merely a pleasant addition to, but an essential part of, the education of the physician. The treatise entitled *That the best doctor is also a philosopher* gives three main reasons for this, one corresponding to each of the three main branches of philosophy, logic, physics and ethics. First, the doctor must be trained in scientific method. The emphasis here is not on the evaluation of evidence, so much as on knowledge of logic, the ability to set out a proof and to distinguish valid from invalid argument. Secondly, it is the task of philosophy to study nature, and the whole of the theoretical side of biology—the investigation of the constituent elements of the body and the functions of the organs, for example—comes under this heading. Thirdly, there is a to us rather surprising ethical reason for the doctor to study philosophy. The profit motive, Galen says, is incompatible with a serious devotion to the art. The doctor must learn to despise money. Galen frequently accuses his colleagues of avarice and it is to defend the profession against this charge that he plays down the motive of financial gain in becoming a doctor. Just as we saw Vitruvius doing for architecture, so Galen attempts as far as possible to assimilate medicine to philosophy, the supreme—because supremely unselfinterested—study.

Many of Galen's interests are poorly represented in the extant works. We know, for instance, that he wrote more than twenty books of commentaries on the logical treatises of Aristotle, but none of these, and very little of his own logical, ethical and philological work, has survived. Even so the extant treatises—which with a Latin translation run to nearly 20,000 pages in Kühn's edition, to which must be added other works that survive only in Arabic translations—are extremely varied both in subject-matter and in style. They cover every branch of the study of health and disease and of 'physiology', the study of the nature of the human body. Some are introductory works, for example his elementary course on anatomy, entitled *On Bones for Beginners*. Others are devoted to the examination of other theorists' views. The commentaries on Hippocrates are a special case, for he represented, for Galen, the repository of medical wisdom. Galen is aware of a 'Hippocratic question', but he is confident that most of the works attributed to Hippocrates in his day were genuine—or that if they were composed by later writers, they nevertheless contained the authentic teaching of Hippocrates. He recognizes that there are gaps in Hippocrates' knowledge, but his general attitude is one of deference to the authority of his great predecessor.

But apart from Hippocrates, other medical and philosophical writers are also discussed at length. Plato holds a place in his admiration second only to Hippocrates himself, and he comments extensively on Aristotle, Herophilus and, especially, Erasistratus, as well as more recent medical theorists such as Asclepiades (first century B.C.) and the founders of the Methodist school, Themison and Thessalus. As with Ptolemy, however, it is a mistake to consider Galen a mere eclectic. Like Aristotle, he quotes his predecessors partly in order to help formulate the problems and partly to establish a body of agreed medical opinion, and he often works out his own view by opposing those of earlier writers. Thus in one of his major theoretical works, *On the Natural Faculties*, he does so by criticizing Erasistratus and the Erasistrateans at considerable length.

Much of his physiology is based on traditional ideas, including, for example, the theory of the four primary elements, fire, air, water and earth, each of which is characterized by two of the four primary qualities, hot, cold, wet and dry. This was, as he knew, Aristotle's doctrine, but he traces it back further still, claiming, on the basis of the evidence in *On the Nature of Man*, especially, that it was Hippocrates who first defined the qualities of the four elements. Other substances are compounds made up of the four elements in differing proportions, and adapting an Aristotelian idea, he distinguishes between homoeomerous, that is homogeneous, parts—such as flesh, bone and blood—and instrumental ones, for example foot and hand.

Thus far Galen was, in the main, simply recapitulating the dominant physical theory of antiquity. But it is important to note how far he was aware of the problems that it presented. Thus he remarks that the primary elements exist only rarely in a pure state in nature. The same applies to the primary qualities too, and he acknowledges how difficult it is to define exactly what is meant by 'hot' 'cold' 'wet' and 'dry'. In *On Mixtures* (book I ch 5) he castigates his predecessors for their failure to do this. For himself, he insists first on the distinction between proper or innate, and acquired, heat, or between things that are hot *per se*, and those that are so *per accidens*. Then among things that are hot *per se* he distinguishes different degrees of heat. First there is what is hot 'simply' or 'absolutely', but only the primary elements are such. Then there are things that are 'predominantly' or 'relatively' hot, 'relative', that is, to some standard with which they are compared, which in the case of an animal, for instance, might be another member of the same species, or other animals. Elsewhere, in his analysis of drugs and their effects (*On Simples*), he classifies hot things into four grades on a scale ranging from the mildest heat (which is not actually perceptible to the senses, but is inferred by reason) to the most extreme—that of burning.

The doctrine of the four primary opposites is fundamental to much Greek physiological and biological theory, yet it was fraught with problems, uncertainties and inconsistencies. Galen recognizes the prevailing confusion and attempts to introduce some order into the discussion. His theory of grades of heat and cold, in particular, represents an attempt to bring quantitative notions to bear. And yet these notions were not applied systematically, nor, in the absence of any objective criterion for measuring heat, were they of any use when they were applied. Even if a thing is not actually hot, it may be potentially so: it may have a heating effect when acting on, or mixed with, some other thing. So Galen warns his reader that things that appear 'hot' may have 'cooling' effects. Neither sensation nor reason is a wholly reliable guide. Yet despite the problematic nature of the theory, he continues to express his confidence that in practice experience will show how the qualities of things are to be determined.

The four primary qualities and the four simple elements form the physical basis of Galen's physiology. But animals possess not only a *physis*, nature, in common with plants, but also *psȳchē*, life or soul. Although he says he does not know what the *substance* of the soul is, he follows Plato (among others) in identifying three main *faculties* of the soul, the rational (*logistikon*), the spirited (*thȳmoeides*) and the appetitive (*epithȳmētikon*), and he correlates these three main vital functions with the three main organs in the body, the brain (the centre of the nervous system), the heart (the source of the arteries) and the liver (thought of as the source of the veins). Like many earlier theorists, he assigns to the liver and the veins themselves the task of making the blood from the nourishment they receive from the stomach and the intestines, and he holds that blood in turn is the matter from which the other parts of the body, including the major organs, are composed. But he draws a distinction between the thicker, denser, muddier venous blood and the thinner, lighter, purer arterial blood which contains the so-called *pneuma zōtikon*, 'vital spirit': he writes obscurely about this being produced in the heart and the arteries from the air we inhale and from 'the exhalation of the humours' (especially blood).[1] Finally, he speaks of another kind of *pneuma*, *pneuma psȳchikon*, the 'psychical spirit',[2] which is derived from the 'vital spirit'—although it is also said to be nourished directly by the air breathed in through the nostrils—and which is located chiefly

[1] He believed that air is drawn into the left ventricle of the heart from the lungs through the pulmonary veins, although it is, in his view, only the quality of the *pneuma*, not its substance, that the heart needs and uses. He also believed (like many Greek theorists) that air is not only inhaled through the nostrils, but also drawn into the arteries through invisible pores in the skin (for example, *On the Natural Faculties*, III ch 14).

[2] The Greek term *pneuma psȳchikon* was translated into Latin as *spiritus animalis* (from *anima*, for Greek *psȳchē*), and this in turn gave rise to the English 'animal spirit', still used by many translators and commentators on Galen. But since 'animal' is potentially a quite misleading rendering of *psȳchikon*, I have avoided it and preferred the transliteration 'psychical'.

in the brain. While the vital spirit is responsible for life itself and the processes essential to it, the psychical spirit is responsible for consciousness and the sensory and motor functions of the nervous system.

His interpretation of the role of the three main organs, liver, heart and brain, is highly schematic, and one of the motives behind the theory is, no doubt, a desire to find a physical vehicle for each of the three main vital functions that he took over from Plato. But although much of the theory is pure speculation, not all of it is as fanciful as is sometimes made out. First, the difference between the bright red arterial, and the darker venous, blood is striking, and to explain this difference —which we see as one between oxygen-rich and deoxygenated blood—in terms of *pneuma* or air is, so far as it goes, on the right lines. Secondly, Galen successfully refuted the common Greek view that the arteries contain nothing but air. As we saw (pp 81 ff), Erasistratus had held this view and had argued that the blood which flows from a cut artery comes not from the artery itself, but from the adjacent veins. But Galen accumulated evidence, for example from experiments on arteries ligatured immediately above and below the point of incision, to demonstrate that the arteries are always full of blood. And he undermined the Erasistrateans' explanation with arguments to show how implausible it was to suppose that there is a sudden evacuation of the *pneuma* from the arteries as soon as one of them is punctured. As for his doctrine of 'psychical spirit', although the idea that this is elaborated in the 'rete mirabile' at the base of the brain is pure conjecture, one point that the doctrine was designed—hopefully—to explain is the obvious fact that an animal deprived of air suffocates and loses consciousness: and Galen's localizing the air responsible for nervous functions in the brain was in part due to the observation that damage to the ventricles of the brain is also followed by loss of consciousness.

Much of Galen's talk of 'natural faculties' is of little value, being no more than a restatement of the phenomena to be explained. Indeed Galen himself comes close to admitting as much at one point in *On the Natural Faculties* (I ch 4) where he says that 'so long as we are ignorant of the substance of a cause which is at work, we call it a *dynamis* (faculty, ability)', giving as instances of this the 'pulsative' faculty of the heart and the 'concocting' or 'digestive' faculty of the stomach. Yet although this looks perilously close to postulating, in the terms of Molière's *Malade Imaginaire*,[1] that the effect of opium is due to a *virtus dormitiva* or 'dormitive faculty', the theory of faculties did not prevent Galen from undertaking detailed investigations of the natural processes in question. One example is his account of digestion in *On the Natural Faculties* III. As usual, much of his discussion is devoted to criticisms of other views. Erasistratus, in particular, had tried to explain nutrition principally in terms of the mechanical action of the gullet and the stomach walls. Galen does not deny the part played by such factors: indeed he describes the results of vivisections that he carried out that confirmed the role played, in digestion, by the peristalsis of the alimentary canal and the contraction of the stomach:

> For I myself have on innumerable occasions dissected the peritoneum of a still living animal and have always found all the intestines contracting round their contents; but the condition of the stomach was not so simple, but when food had just been taken, it grasped this exactly above and below and on all sides and was without movement . . . At the same time I found the pylorus always shut and closed exactly like the mouth of the womb in pregnant women. But where the digestion had been completed, the pylorus had opened and the stomach was moving peristaltically in the same way as the intestines. (III ch 4.)

But he differs from Erasistratus in insisting that other factors must also be taken into account. Food is not only driven along the alimentary canal: the stomach exercises an attraction, *holkē*, upon it. More important, the idea of the specific attraction of certain substances for one another is used to explain the assimilation of nourishment. Nutrition involves much more than a simple mechanical process: the food must first be emulsified or turned into chyle (*chȳlōsis*), then undergo concoction or digestion (*pepsis*) and finally be

[1] This play, first produced in 1673, contains much gentle mockery of contemporary medical ideas, which remained deeply influenced by ancient medical thought in general and by Galen in particular.

absorbed (*anadosis*), this last being a process in which like substances, or those between which an affinity exists, are assimilated to one another.

Although much of Galen's physiology is highly speculative, his descriptions of both anatomical structures and physiological processes show evidence of sustained and minute observation. The texts that imply that human dissection continued on only a very limited scale in Galen's day have been reviewed in chapter 6 (pp 86 ff). From these it is clear both that Galen's ideal is to perform dissections on human subjects, and that the opportunities he had to do so were rare. As a second best, dissection should be carried out on animals, and this is indeed recommended in any case as an addition to, and preparation for, human dissection. 'I want you to have frequent practice on them [animals like men]', he says in *On Anatomical Procedures* III ch 5, 'so that if you have the luck to dissect a human body, you will be able readily to lay bare each of the parts'. For his own studies, Galen used a variety of animals, including pigs and kids,—on one notable occasion he even dissected an elephant,—but for preference he worked with apes, telling his students to

> choose those apes that are most like man, such as those whose jaws are not protruding and in which the so-called canines are not large. In such apes you will find the other parts too arranged like man's and for this reason they walk and run on two legs. . . . Those that are most like man very nearly have a completely erect posture. For the head of the femur fits into the socket at the hip joint rather obliquely and some of the muscles that extend towards the tibia go further [than in man]. (*On Anatomical Procedures* I ch 2.)

Much of the work that went into his anatomical masterpiece, *On Anatomical Procedures*, was done on the barbary ape. This certainly led Galen into errors concerning human anatomy, although it is clear from several passages including the one just quoted that he was at least to some extent aware of the danger of arguing by analogy from apes to man.

Galen insists, above all, on *practice* in dissection. Just as the great sixteenth-century anatomist Vesalius was to do, he castigates arm-chair professors who did their anatomy out of books. Even in his day this habit had evidently begun, and it was, of course, to spread once Galen's own works became the text-books from which the professor lectured. Time and again, and in other contexts besides the study of anatomy, he emphasizes the inadequacy of book-learning by itself, even though knowledge of the great doctors of the past is an essential part of medical training. He warns his students, too, of the dangers of having an assistant to do part of the work. 'At first,' he says in *On Anatomical Procedures* I ch 3,

> I too had an assistant to skin the apes, avoiding the task as beneath my dignity. Yet when one day I found by the armpit, resting on and united to the muscles, a small piece of flesh which I could not attach to any of them, I decided to skin the next ape carefully myself. I had it drowned, as I usually do, to avoid crushing the neck, and tried to remove the skin from the surface, avoiding the organs beneath. I then found, extended under the whole skin of the flank, a thin membranous muscle . . . Having found this muscle, the nature of which will be fully and duly explained, I was the more anxious to skin the animals myself, and thus I discovered that nature had wrought these aforesaid muscles for important functions.[1]

The techniques of dissection, and still more vivisection, are difficult, and can only be acquired by long training and practice. He several times remarks how hard it is to perform a dissection successfully at the first attempt; it is only by constantly practising dissection that familiarity with anatomical structures can be gained, for 'no phenomenon is accurately and quickly recognized unless often seen'; and he observes that it was only by repeating the same dissection several times that some of his own discoveries were made. A further passage on the dissection of the muscle leading into the armpit illustrates his insistence on care and precision:

> in passing laterally to the false ribs, if you are careless you may tear away the head of the small muscle which, I said,

[1] From the translation in C. Singer, *Galen, On Anatomical Procedures*, Oxford University Press, 1956.

runs into the armpit and was unnoticed by the anatomists . . . This runs up to the armpit, where its fibres converge into a narrow fleshy strand. If you strip away its expanded lower origin with the skin, you will find that the fleshy part extended to the armpit is rent. If on the one hand you are diligent and seek the point from which it is torn and do not find it, you will be full of doubt, as I was at first. But on the other hand, if you are careless and easy-going (as our anatomical predecessors demonstrably were in many of their operations), holding this fleshy sheet to be of no account, you will cut or tear it away from the underlying tissues and throw it away. As to the need for exercising precision in removing the skin there, enough has now been said. (V ch 7.[1])

But in some cases dissection is not sufficient for Galen's purposes. To investigate the vital processes, vivisection must be performed, although he recommends that, before proceeding to dissect the live animal, the operations should be practised on a dead one. One example, the investigation of the processes of digestion, has already been mentioned (p 142). In *On Anatomical Procedures* VII ch 12 ff he describes the vivisection of the heart and lungs which he carried out first to observe the beating of the heart and arteries and secondly to study the effect of constricting the heart artificially. He refers to the difficulties presented by the operation, and particularly to that of controlling the blood:

Nothing upsets any operation like haemorrhage. Bearing this in mind, immediately you see blood spurt from the artery with the downward incision, turn the lancet as quickly as possible to the transverse incision. Then with the thumb and index of the left hand, grasp that part of the sternum where the artery is pouring forth blood, so that while the one finger acts as a stopper for the orifice, both grasp the bone.[2]

[1] From the translation in C. Singer, *Galen, On Anatomical Procedures*, Oxford University Press, 1956.

[2] From the translation in C. Singer, *Galen, On Anatomical Procedures*.

In other vivisections the element of active interference with nature—the creation of artificial conditions to investigate vital functions—is more pronounced. Thus in *On the Natural Faculties* I ch 13 he describes vivisections used to demonstrate the function of the ureters. Against Asclepiades, who had argued that the urine passes into the bladder by being dissolved into vapours, he shows first that the bladder of a living animal is filled through the ureters, and that if the ureters are ligatured, no urine passes into the bladder, but the ureters themselves become distended with urine. And then he proceeds to demonstrate that the direction of this flow is irreversible: if the base of the bladder is ligatured and one attempts to force urine back up the ureters to the kidneys, the ureters prevent this, acting as a valve and stopping any fluid passing back up them.

Even more remarkable are the vivisections that he performed to investigate the nervous system. In these he conducted a series of tests in which he made incisions either right through the spinal cord, or through one side of it, at various points on the spinal column, in order to discover what effect this had on the animal's faculties. He describes the vivisections in *On Anatomical Procedures* IX ch 13 f:[1]

> Accordingly commence at the meeting-place of the so-called 'greatest' bone . . . [sacrum] with the last vertebra, where a massive nerve-shoot grows outwards so as to distribute itself to the lower limb. Next, go upwards to the vertebra above this, then further upwards. And on the several [nerves, one by one] you will see . . . that the first structures which paralysis affects and which will be deprived of movement, are the ends of the legs, the second . . . the parts which come in front, then the parts of the thigh and the hips, then those of the lumbar region. When you come to the thoracic vertebrae, then the first thing that happens is that you see that the animal's respiration and voice have been damaged.[2]

[1] From the translation in W. L. H. Duckworth, *Galen, On Anatomical Procedures, The Later Books*, Cambridge University Press, 1962.

[2] In connection with Galen's work on the voice, it may be noted that one of his earliest anatomical discoveries was that of the recurrent

He continues systematically up the spinal column. Thus:

> Should the cut follow a line behind the second thoracic rib, then that does not damage the arm, except that the skin of the axillary cavity, and the first subdivisions of the region of the upper arm turned towards the trunk become deprived of sensibility.

And again:

> Transection of the spinal marrow behind the fifth [cervical] vertebra paralyses all the remaining parts of the thorax, and arrests their movements, but the diaphragm remains almost unscathed.

Despite the lack of effective anaesthetics and of reliable antiseptic agents, he evidently pursued his investigations with great skill and persistence. This is not to deny, of course, that much of his anatomical work is inaccurate. Some of his mistakes come from a failure to distinguish carefully enough between animal, and human, structures. More often they reflect inadequate or erroneous physiological notions or they stem from an overeagerness to establish the uses of parts and thereby demonstrate the purposefulness of nature.

One case where he was misled by faulty argument as much as by faulty observation is in what is, perhaps, his most notorious error, the belief that blood is carried over directly from the right into the left side of the heart through the interventricular septum, the solid musculo-membranous partition that divides the two sides of the heart. Galen knew that blood is carried into the right side of the heart by the Vena Cava, and he was well aware of the action of the four main valves of the heart, each of which serves as a one-way valve.[1] Thus the tricuspid valve allows blood to enter the right ventricle but prevents it from regurgitating into the right atrium and the Vena Cava. He knew too that there was no

---

laryngeal nerves, of which he gives an exact description in *On Anatomical Procedures* XI ch 3 ff.

[1] The mitral valve at the entrance of the left ventricle is, however, in his view, a less effective valve than the other three, in that it allows some matter described as 'sooty residues' to pass out of the left ventricle back into the pulmonary veins.

question of the blood that had entered the right ventricle being used up in it. The heart muscle itself is supplied with blood by the coronary blood-vessels. Given that there was blood in the left side of the heart (and we have seen that he refuted Erasistratus' idea that the arteries and left heart contain nothing but air) he was faced with the problem of

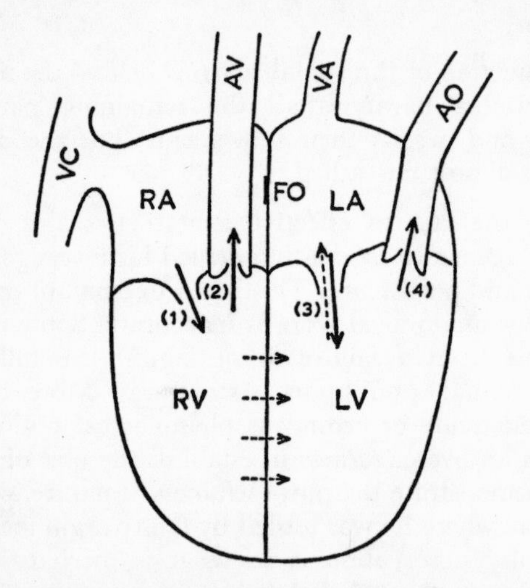

Fig. 34 Schema to represent blood flow in the heart according to Galen. The arrows indicate the direction of flow through the four main valves of the heart. As indicated by the broken arrows, he believed that blood passes directly from the right to the left ventricle through the interventricular septum, and that there is some reflux from the left ventricle to the venous artery through the mitral valve.

Key:
- VC Vena Cava
- RV Right Ventricle
- LV Left Ventricle
- VA 'Venous artery' (i.e. pulmonary vein)
- AO Aorta
- RA Right Atrium
- LA Left Atrium
- AV 'Arterial vein' (i.e. pulmonary artery)
- FO Foramen ovale (open in the foetus but closed after birth)

(1) Tricuspid valve
(2) Semi-lunar ('sigmoid') pulmonary valve
(3) Mitral valve (two cusps)
(4) Semi-lunar ('sigmoid') aortic valve

how it got there. Although he does not describe the lesser, pulmonary circulation in so many words, he appreciated that the pulmonary artery takes blood from the right heart into the lungs, and that blood from the lungs finds its way back to the left heart through the pulmonary veins. Moreover he knew that *in the foetus* blood is carried over directly from the Vena Cava into the left atrium through the opening between the atria known as the foramen ovale. This is clearly described in book VI ch 21 of *On the Use of Parts*, and later in the same work (XV ch 6) he refers to the closing of the foramen after birth.

But two main considerations led him to infer that some blood must also pass directly from the right to the left heart through the interventricular septum after birth. First he was misled by the apparent difference in calibre of the openings allowing blood into and out of the right ventricle. He thought the tricuspid orifice (allowing blood to enter) much larger than the pulmonary artery and argued from this that the latter could not be the only way by which the blood leaves the right ventricle. Secondly, he argued by analogy with what happens elsewhere in the body. Following Erasistratus, he believed, correctly, that the arterial and venous systems are interconnected through minute, invisible *anastomōseis*, or capillaries, all over the body. Requiring a route for the blood to pass from the right to the left heart, he argued that similar minute passages must exist within the septum, even though he admitted that their existence could not be directly verified by observation. 'Similarly also, in the heart itself,' he writes in *On the Natural Faculties* (III chapter 15),

> the thinnest portion of the blood is drawn from the right ventricle into the left, owing to there being perforations in the septum between them: these can be seen for a great part [of their length]; they are like pits with wide mouths and they get constantly narrower; it is not possible, however, actually to observe their extreme terminations, owing both to the smallness of these and to the fact that when the animal is dead all the parts are chilled and shrunken. Here, too, however, our argument, starting from the principle that nothing is done by nature without an aim, discovers these anastomoses between the ventricles of the heart; for it

> could not be at random and by chance that there occurred pits ending thus in narrow terminations.[1]

The reference to the principle that nature does nothing without an aim takes us to the question of the motives that prompted Galen's inquiry. Several texts in *On Anatomical Procedures* are explicit on this point and they make it clear that there are, in his view, several possible reasons for studying anatomy. In book II ch 2, for instance, he discusses the usefulness of anatomy under four separate heads:

> Anatomical study has one use for the natural scientist [*anēr physikos*] who loves knowledge for its own sake, another for him who values it not for its own sake but rather to demonstrate that nature does nothing without an aim, a third for one who provides himself from anatomy with data for investigating a function, physical or mental, and yet another for the practitioner who has to remove splinters and missiles efficiently, to excise parts properly, or to treat ulcers, fistulae and abscesses.[2]

He goes on to remark (chapter 3) that in general anatomists have concentrated on 'that part of anatomy that is completely useless for physicians or that which gives them little or only occasional help', instancing the study of the heart and the blood-vessels communicating with it. Many anatomists who studied such matters were ignorant of more mundane, but more useful anatomical knowledge:

> The most useful part of the science of anatomy lies in just that exact study neglected by the professed experts. It would have been better to be ignorant of how many valves there are at each orifice of the heart, or how many vessels minister to it, or how or whence they come, or how the paired cranial nerves reach the brain, than [not to know] what muscles extend and flex the upper and lower arm and wrist, or thigh, leg and foot, or what muscles turn each of these laterally, or how many tendons there are in each . . .

[1] Based on the Loeb translation of A. J. Brock, *Galen, On the Natural Faculties*, Cambridge, Mass., Harvard University Press; London, Heinemann, 1916.

[2] Based on the translation in C. Singer, *Galen, On Anatomical Procedures*, Oxford University Press, 1956.

> or where a vein or a great artery and where a small underlie them.

As he puts it elsewhere (ch 2):

> What could be more useful to a physician for the treatment of war-wounds, for extraction of missiles, for excision of bones . . . than to know accurately all the parts of the arms and legs . . . If a man is ignorant of the position of a vital nerve, muscle, artery or important vein, he is more likely to be responsible for the death, than for the saving, of his patients.

Galen himself was clearly motivated by both the main types of reasons that he mentions in this passage. As a practising physician and ex-surgeon to gladiators, he was well aware of the practical applications of anatomical knowledge. At the same time he shares the motives of the man who values anatomy in order to show that nature does nothing without an aim, and of the 'natural scientist' (in so far as he is to be distinguished from the former) who loves knowledge for its own sake. He is keen to redress the balance against those who had purely theoretical interests in mind and who were thereby led to underestimate the practical aspects of the study. But his own views are inclusive and both types of consideration weigh with him. Just how important the belief in the purposefulness of nature was to him comes out in a revealing passage in *On the Use of Parts*, a work in seventeen books devoted to showing the useful function that each of the parts of the body serves. He several times speaks of the study of anatomy in terms drawn from the mystery religions, as an initiation (*teletē*) into the mysteries (*mystēria*) of nature (for example, book VII ch 14), and in book III ch 20 he calls his work

> a sacred book which I compose as a true hymn to him who created us: for I believe that true piety consists not in sacrificing many hecatombs of oxen to him or burning cassia and every kind of unguent, but in discovering first myself, and then showing to the rest of mankind, his wisdom, his power and his goodness.

Galen's four-fold justification of anatomy suggests that the study needed defending in his day, no less than the practice of dissection had done in that of Aristotle.[1] Indeed the circumstances in which anatomical problems were discussed were sometimes far from conducive to the calm pursuit of knowledge. The profession that provided Galen with a livelihood was, as I have said, an extremely competitive one, and even in the study of anatomy the element of rivalry is present. As doctors often debated the diagnosis of a case in front of the patient, so points of dispute on questions of anatomy were sometimes settled at public demonstrations. One case he refers to is that of an Erasistratean who was challenged to show an artery empty of blood (*On Anatomical Procedures* VII ch 16). First he said he would not do so without a fee, whereupon the bystanders put down a thousand drachmae for him to pocket if he succeeded. Galen continues: 'in his embarrassment he made many twists and turns, but, under pressure from all present, mustered courage to take a lancet and cut along the left side of the thorax especially at the point where, he thought, the aorta should become visible. He proved so little practised in dissection that he cut on to the bone.'[2]

In medicine, too, Galen describes his colleagues as descending to every kind of corrupt practice to build up their own reputations and to undermine those of rivals, and this helps to explain the strident, boastful tone of many of his stories about his own work. Although his views on the causes and treatment of disease are, in the main, strongly traditional, he seldom admits to giving an incorrect diagnosis and even more rarely to failing to produce a cure. He reports that his own skill as a diagnostician gained him a name for magical powers. Even though the use of the pulse in diagnosis goes back to Praxagoras and Herophilus (see above pp 79 f), Galen had to defend his own use of it against the charge that he was practising divination (*mantikē*). He often insists—much in the same way that the Hippocratic writers had done—on the difference between the doctor and the soothsayer, between the doctor and the mere drug-seller (*pharmakopōlēs*) or more simply between the doctor and the layman. Yet we find him maintaining the value of dreams in diagnosis, and he acknowledges,

[1] See *Early Greek Science*, p 105.

[2] From the translation in C. Singer, *Galen, On Anatomical Procedures*, Oxford University Press, 1956.

though he likes to minimize, the element of guesswork in both diagnosis and treatment. Much research in the biological sciences was done by doctors. Yet it is clear not only that different justifications—practical and philosophical—were offered for that research, but also that what the rational art of medicine itself should be taken to include continued to be as much a matter of dispute in the second century A.D. as it had been in the fifth century B.C.

10

The Decline of Ancient Science

THE topic of when and why ancient science began to decline is, like so much else in our study, as controversial as it is complex. An extreme view was expressed by Cornford, for example, when he wrote that 'all the most important and original work (in the inquiry into the nature of things) was done in the three centuries from 600 to 300 B.C.'[1] Even after 200 B.C., at the end of an exceptionally creative century, important work continued to be done. In astronomy, for instance, there is Hipparchus in the second century B.C., and although few names of note are known to us between him and Ptolemy, there are references in the *Almagest* to observations carried out in the first century A.D. by Agrippa of Bithynia and by Menelaus of Alexandria, who was also responsible for a treatise on the geometry of the sphere, the *Sphaerica*, that is important for the history of trigonometry. In the history of medicine and biology, evidence of active investigations during the period between Erasistratus and Galen comes not only from the pages of Galen himself, but also from the extant writings of Rufus and Soranus. Any thesis that Greek science ground to a halt in the third or second century B.C. has, in any case, to take Ptolemy and Galen into account, both of whom must be considered among the greatest scientists that the ancient world produced.

Ancient science is, we must repeat, made up of the contributions of authors of many different types, philosophers, mathematicians and astronomers, doctors, even architects and engineers. To talk of the decline of ancient science in global terms, ignoring these distinctions, is as potentially misleading as it is to do so of its origins and early growth. Our first task is to consider briefly what signs there are of continued inquiry, and in particular what evidence there is of original scientific thought,

[1] 'Greek natural philosophy and modern science' in F. M. Cornford, *The Unwritten Philosophy and other essays*, Cambridge University Press, 1950, p 83.

after 200 A.D., in each of the three main branches of the investigation concerning nature, namely (1) natural philosophy (including physics and cosmology), (2) mathematics and astronomy and (3) biology and medicine. The histories of these subjects after the end of the second century A.D. differ in certain respects, but in each case the hypothesis of an abrupt cessation of inquiry is untenable.

Take first natural philosophy. By the end of the second century the great Hellenistic philosophies of Epicureanism and Stoicism had for some time been in decline. Although there continued to be men who called themselves Epicureans or Stoics, neither school produced any major thinker after 200. The Epicureans, in any case, aimed to preserve unchanged the thought of their founder, and the last important Stoic philosopher was the emperor Marcus Aurelius (121–180 A.D.). Yet while these schools declined, others grew. The third century saw a rebirth of Platonism, and in Plotinus (205–270) neo-Platonism found an original philosopher of a very high order. His chief interests were theological and metaphysical, and his importance for the history of physical speculation is small. But later neo-Platonists, especially Iamblichus and Proclus, were responsible for a revival of aspects of Platonism that are relevant to physics.

Iamblichus, a Syrian by birth who lived at the turn of the third and fourth centuries is a fascinating combination of mystic and rationalist—though in this no more extraordinary than the Presocratic philosopher Empedocles, for example. Iamblichus' belief in theurgy—the working of things divine—comes out especially in *On the Mysteries*. On the other hand, in his *On the Common Mathematical Science*, he explores the applications of mathematics to science.[1] He remarks (ch 23) that the Pythagoreans valued mathematics and applied it in many different ways to the study of the cosmos: 'they considered what is possible and impossible in the structure of the universe on the basis of what is possible and impossible in mathematics, and they apprehended the celestial revolutions with their causes according to commensurate numbers', and he suggests (ch 32)

[1] Similar ideas had been expressed by neo-Pythagoreans such as Nicomachus of Gerasa, who wrote an influential *Introduction to Arithmetic* towards the end of the first century A.D.

that mathematics provides the key to the understanding not only of the movements of the heavenly bodies, but of natural phenomena as a whole:

> It is also the custom of mathematics sometimes to attack mathematically perceptible things as well, such as the four elements by using geometry or arithmetic or harmonics, and similarly with other matters. For since mathematics is prior in nature, and is derived from principles that are prior to those of natural objects, for that reason it constructs its demonstrative syllogisms from causes that are prior . . . Thus I think we attack mathematically everything in nature and in the world of coming-to-be.

Here Iamblichus goes far beyond Plato, indeed beyond any ancient writer, in advocating the mathematization of the whole of the study of nature. Admittedly this remained merely an ideal, and he is prepared to cite almost any example of the application of numbers to perceptible phenomena to support his general thesis. But even though the idea of 'attacking mathematically everything in nature' had to wait until modern times to be realized, and it was then realized in a way that would have astonished Iamblichus, the importance of this passage as the most explicit statement of this ideal in any ancient writer remains.

Proclus of Byzantium, who became head of the Platonic Academy at Athens in the late fifth century A.D., was, like most of the neo-Platonists, first and foremost a metaphysician. But in a lost work the arguments of which are reported by Simplicius he undertook a defence of Plato's geometrical atomism against the criticisms of Aristotle and others. Thus it had been objected against Plato that his theory did not allow earth to change into the other simple bodies, because earth consists of triangles of a different kind from those that constitute the other three elements.[1] In meeting this difficulty, Proclus insists on the distinction between pure earth and earthlike substances. The former, he argues, is not seen to change into other elements: pure earth as such is unchangeable. But 'earthlike substances change in so far as they are contaminated with air and water'. But he goes beyond Plato's

[1] See *Early Greek Science*, pp 74 f.

explicit teaching particularly in his analysis of the changes that take place between the other three elements, where he allows—what Plato had not, at least not expressly—that the triangles that constitute the simple bodies may exist temporarily in suspension. The change from air to water, for instance, takes place in two stages, and in the process of being synthesized to form icosahedra of water, some of the triangles from the original octahedra of air exist momentarily *neither* as fire *nor* as air.

On these and other points Proclus clarifies and modifies Plato's doctrines. His primary motive is to interpret and defend Plato. But in so doing he goes beyond exegesis and enters the debate between the quantitative and the qualitative theories of matter with arguments of his own. Although he brings no new empirical evidence to bear on the problem, his work is not a repetition, but an elaboration, of Plato's theory.

Evidence for the continued discussion of physical problems in the next century comes from two of the commentators on Aristotle, the pagan neo-Platonist Simplicius of Athens, and the Christian John Philoponus of Alexandria. Simplicius' commentaries on the *Physics* and *On the Heavens* are written in a spirit of deference to Aristotle's authority on most problems of physics. However he provides very full information concerning earlier and contemporary arguments both for and against Aristotelian doctrines and it is clear that he investigated some problems on his own account. One example is the question of whether air has weight in its own place. In *On the Heavens* (311 b 8 ff) Aristotle had stated that 'in its own place everything has weight except fire, even air' and had adduced as proof of this that an inflated bladder weighs more than an empty one. In his comment on this passage (710 14 ff) Simplicius mentions that Ptolemy had contradicted Aristotle and had sought to prove that air has no weight in its own medium by the same experiment of the inflated bladder. 'Ptolemy not only contradicts Aristotle's view that the inflated bladder is heavier than when uninflated, but he maintains that the inflated bladder actually becomes lighter.' Simplicius then goes on: 'Having tried this out myself with the greatest possible accuracy I found that the weight of the bladder when uninflated and inflated was the same.'

In this case the test had been suggested, and, it seems, carried out, before him. Like many ancient experiments, it led to no decisive result.[1] But the text illustrates that Simplicius not only attempted to adduce fresh arguments on the problems referred to in the works of Aristotle, but also was capable, on occasions, of investigating those problems empirically himself.

Whereas Simplicius generally wrote in support of the Aristotelian position, his slightly older contemporary John Philoponus engaged on a sustained two-pronged attack on Aristotelianism, criticizing many of the central theses of Aristotle's physics both on the grounds of inconsistency and on the grounds that they fail to fit the facts. He too draws extensively on earlier work, particularly that done by Strato and Hipparchus. But while Philoponus was certainly not the first writer to attack some of the fundamental theses of Aristotelian physics, his is by far the most comprehensive extant ancient refutation of them.

His interests included cosmology, theology, logic and the whole range of 'the inquiry concerning nature', and his work in one field, dynamics, is outstanding. The effectiveness of his arguments can be judged by considering his criticisms of 'antiperistasis'. This is the doctrine that a projectile is kept in motion by the pressure of air behind it, the idea being that the air displaced in front of the projectile moves round and pushes the projectile from behind.[2] First Philoponus objects that no reason was given for the air to come round behind the missile at all. Taking the case of an arrow shot from a bow, he asks:

> How is it, then, that the air, pushed by the arrow, does not move in the direction of the impressed impulse, but instead, turning about, as by some command, retraces its

[1] The different results obtained by Aristotle, Ptolemy and Simplicius may not be due simply to negligence in carrying out the test. The weight of the inflated bladder depends on, among other things, the proportion of carbon dioxide it contains, and this will vary according to whether it is filled with exhaled breath or atmospheric air.

[2] Aristotle himself had rejected the doctrine in this form but he still held that in forced motion the power of causing movement is imparted by the projector to the medium (air or water) behind the projectile.

course? Furthermore, how is this air, in so turning about, not scattered into space, but instead impinges precisely on the notched end of the arrow and again pushes the arrow on and adheres to it? Such a view is completely implausible and is more like fiction. (*Commentary on Aristotle's Physics*, 639 30 ff.[1])

Later on he asks whether the initial impetus imparted to a missile is to be explained in the same way:

> When one projects a stone by force, is it by pushing the air behind the stone that one compels the latter to move in a direction contrary to its natural direction? Or does the thrower impart a motive force to the stone too? If he does not . . ., of what advantage is it for the stone to be in contact with the hand, or for the bowstring to be in contact with the notched end of the arrow? (641 13 ff.)

This leads him to consider an imaginary experiment in which an arrow or a stone is set up on a stick and an attempt is made to move it by setting a large quantity of air in motion behind it 'with countless machines':

> The fact is that even if you place the arrow or stone upon a line or point quite devoid of thickness and set in motion all the air behind the projectile with all possible force, the projectile will not be moved the distance of a single cubit. (641 23 ff.)

The important positive conclusion that Philoponus reaches is that so far from the medium being necessary for motion—as the Aristotelians had assumed—the effect of the medium is to resist the motion. Forced motion cannot be caused by the pressure of the air behind the projectile. On the contrary, 'it is necessary to assume that some incorporeal motive force is imparted by the projector to the projectile'. And he correctly concludes that there is nothing to prevent the transmission of an impetus taking place in a void. 'It is quite evident that if one imparts motion contrary to nature or forced motion to an

[1] Based on the translation in M. R. Cohen and I. E. Drabkin, *A Source Book in Greek Science* (second edition), Cambridge, Mass., Harvard University Press, 1958.

arrow or a stone, the same degree of motion will be produced much more readily in a void [than in a plenum].'

The refutation of 'antiperistasis' involves a thought experiment. His discussion of the Aristotelian doctrine of the relation between the weight and the speed of a freely falling body refers to actual tests that he appears to have undertaken himself. Aristotle's view, he suggests, implies that in motion through the same medium the times required for the movements will be in inverse proportion to the weights of the moving bodies. 'But,' he says (683 16 ff), 'this is completely false, and this can be established by what is actually observed more powerfully than by any sort of demonstration by arguments', and he then proceeds to adduce (as a thousand years later Galileo was to adduce) the evidence of what happens when two different weights are dropped from the same height:

> For if you let fall at the same time from the same height two weights that differ greatly, you will see that the ratio of the times of the motions does not correspond to the ratio of the weights, but that the difference in the times is a very small one.

Philoponus carried out a devastating critique of Aristotelian dynamics. Where Aristotle had insisted on the difference in kind between 'natural' and 'forced' motion, Philoponus undermined that distinction. Whereas Aristotle had argued that the medium is necessary for motion, Philoponus showed that the medium acts purely as a resistance to the moving object. How much his impetus theory owes to the ideas of Hipparchus is hard to judge, but Philoponus appears to have stated more clearly than any earlier writer that what is imparted to the moving object is an *immaterial force*. Moreover he supported his theoretical arguments with the data of observation and deliberate experiment, and he made explicit his preference for empirical evidence over verbal argument.

Yet although he exposed some of the errors in Aristotle's dynamical theories, he did not succeed in putting dynamics on a firm footing. In particular, he failed to realize that in a void the speed of bodies is not affected by their weight. He held, correctly, that weight is only one of the factors that govern the speed of a body falling through a plenum. Against Aristotle he argued that the speed is not inversely proportional to the density of the medium, but is made up of two components, one that varies with the weight of the moving body, and a second that varies with the density of the medium. But whereas Epicurus, for example, had held that in motion through a void speed does not vary with weight, Philoponus rejected this view and argued, this time incorrectly, that in a void weight would continue to be one of the variables on which the speed of a falling body depends.

We have no space here to discuss Philoponus' work in other branches of physics, but a word must be said about some of his more general cosmological doctrines. Once again the starting-point of his own speculations is often the criticism of Aristotle, for example, of the doctrine of the *aithēr* and the distinction between the heavenly and the sublunary region—where Aristotle's view had already been called in question by, among others, his immediate successor, Theophrastus. On this problem too Philoponus' attack combines theoretical and empirical considerations. Thus he claims that astronomy had shown that the movements of the planets are not simple, and he infers from the observed differences in the colours of the planets that there must be differences in their constitution. But if the substance of the heavenly bodies exists in different forms, then on good Aristotelian principles it follows that the heavens cannot be immune to change.

Philoponus thus denies any difference in kind between the heavenly and the sublunary sphere. The heavens are not composed of a separate fifth element, which has a special natural movement in a circle, but rather of fire. Whereas Aristotle had denied that hot, cold, dry and wet exist in the heavens, Philoponus insists that they do. 'As it is visible,' he is quoted by Simplicius as arguing, 'it is assuredly also tangible, and being tangible it possesses tangible qualities, hardness, softness, smoothness, roughness, dryness, wetness and the like, and heat and cold which include all these.' Admittedly he is influenced, just as Aristotle and the later Aristotelians had been, by theological considerations: one of his arguments is based on his view that God alone is omnipotent. But he answers physical arguments with physical arguments of his own, exposing the inconsistencies in Aristotle's doctrine and powerfully advocating the contrary view of the essential unity of the universe.

The examples we have considered show that cosmology and some branches of physical inquiry were kept alive long after the second century A.D. Although much effort was expended on commenting on earlier theorists, that did not prevent new ideas from being expressed. The neo-Platonists rediscovered and extended the notion of the mathematization of science, and Philoponus' criticism of Aristotelianism was not merely derivative, and not only destructive, but in parts profoundly original and creative, even if it offered no new comprehensive physical system.

The history of mathematics and astronomy after the second century follows a roughly similar pattern. Again the commentary and the outline history provide the main vehicles of thought. Euclid's *Elements*, especially, was the subject of numerous editions and commentaries, but the scholars who composed these (who included Pappus of Alexandria in the early fourth century, Theon of Alexandria in the late fourth, Proclus in the fifth and Simplicius in the sixth) not only undertook the exegesis of the text, but also on occasions introduced new ideas and arguments of their own. Thus Proclus was one of those who attempted to construct a proof of Euclid's fifth postulate concerning parallel lines (see above, pp 37 f). But more important, some original mathematical work was also done outside the commentaries and histories. One outstanding work that does not take the form of a commentary on an earlier text is the *Arithmetica* of Diophantus of Alexandria, who probably lived about the middle of the third century. This work stands in a similar relation to the history of Greek algebra as Euclid's *Elements* to Greek geometry. The *Arithmetica* includes much that was known long before the third century A.D., but Diophantus not only systematized earlier knowledge, but also added much to it.

Astronomy fared worse than mathematics. Advanced astronomy, including much of Ptolemy's *Almagest*, was, quite simply, too advanced for the type of popular handbook produced in late antiquity. Commentaries on, or outlines of, the *Almagest* were, however, written by Pappus and Theon, and Proclus composed an *Outline of the Astronomical Hypotheses*—

an introduction to astronomical theory—as well as a paraphrase of, and perhaps also a commentary on, the *Tetrabiblos*. It is clear that astronomical observation continued to be practised. Indeed a good deal of space is devoted in Proclus' *Outline* to describing how to make various astronomical instruments, and Philoponus composed a treatise on the construction of the plane astrolabe. Yet such observations as were made were undoubtedly more often undertaken either for practical purposes, such as the regulation of the calendar, or in connection with astrological beliefs, than in the hope of advancing the investigation of problems in astronomical theory. The prevailing mood is one of resignation—fostered in part no doubt by Ptolemy's own admissions of the difficulties of the problems (see above pp 128 f): there is a sense of the hopelessness of attempting to discover the truth about the motions of the heavenly bodies. Thus although Proclus gives a clear description of Ptolemy's system, he comments on current astronomical theories that 'these hypotheses do not even have any probability, but some are far removed from the simplicity of divine things, and others, fabricated by more recent astronomers, suppose the motion of the heavenly bodies to be as if driven by a machine' (*Commentary on the Timaeus* III 56 28 ff). Again in his work *On the Construction of the World* (III 3) Philoponus sets out to show that the hypotheses of the astronomers are unproved, and that nobody has succeeded, nor ever will succeed, in giving a proof of them. Then in the next chapter, after alluding to the different periods of revolution of the heavenly bodies and to the precession of the equinoxes, he says:

> Who would be able to state the cause of these things? No more could any human being give an account of the number of the stars, their position and order, and the difference in their magnitudes and colours. This only we can say, that God has made everything well and as is needed, neither more nor less. Altogether we know the causes of few things. If therefore people cannot give the natural cause of things that are apparent, they should not keep asking for the cause of things that are not apparent.

Finally I turn to biology and medicine. In one respect these were in a less favoured position than mathematics or astronomy.

Arithmetic, geometry and elementary astronomy formed, with music, the 'quadrivium', which together with the 'trivium' (grammar, logic and rhetoric), constituted the fundamental curriculum of Greco-Roman education even before those terms were coined to describe them. But in another respect medical studies were better placed than other branches of science. Doctors were always needed, and there is no suggestion that the medical profession declined in numbers or in prestige after Galen. The medical schools continued to flourish and they insured that a good deal of medical knowledge was handed on from one generation to the next.

As in astronomy and other fields, an increasing proportion of medical writings takes the form of commentaries, digests or histories after the second century, and Galen soon takes his place alongside Hippocrates as the model of the physician. In the mid-fourth century, for instance, Oribasius (born like Galen at Pergamum) composed a vast *Medical Collection*, a medical encyclopedia in 70 books about one third of which is extant. He explains the plan and purpose of the work in the preface addressed to his friend and patron, the Emperor Julian:

> When, having praised my first compendium, you commanded me to make a second, by searching out and collecting all that is most important from all the best doctors, and all that contributes to the goal itself of medicine, I eagerly agreed to do this as well as I could, believing that such a collection would be very useful . . . I shall make my collection only from the best authors, not omitting any of the material I originally obtained from Galen alone, on the grounds that he is supreme among all those who have written about the same subject since he uses the most exact methods and definitions, as one who follows the Hippocratic principles and opinions.

A similar intention lies behind the work of other later medical writers. In the sixth century Aetius of Amida and Alexander of Tralles composed similar compilations, Alexander's being the more original of the two, and so too in the seventh did Paul of Aegina. Indeed Paul refers to Oribasius' work in the proem to his own treatise, where he states that he considered it too bulky and so made a shorter and more convenient compendium of his own.

The main effort of most medical writers after Galen about whom we have definite information is towards summarizing and systematizing medical knowledge, and their summaries tend to become more concise as time goes on. Yet it would be false to conclude that medical learning became purely and simply book learning. The compilers of handbooks were medical practitioners as well as scholars. There is, too, some evidence that dissection was occasionally practised, although for the purposes of medical education rather than for those of original research. Thus Theophilus Protospatharius, whose date is uncertain but who is not earlier than the seventh century, frequently exhorts his readers both to consult the 'anatomists who practise dissection' and to carry out dissections for themselves, indeed to do so 'often and on many specimens' (*On the construction of the human body*, IV 1, V 4 and 11). To be sure, dissection is used merely to *confirm* the facts, and he does not suggest that it is to be used as a method of research, to make *new* discoveries. Yet it appears that Theophilus is not merely repeating what he had read in Galen, and that dissections may have continued to be undertaken, from time to time, after the sixth century.

A rapid glance at the principal evidence is enough to dispose of the idea that scientific inquiry in all its forms ceased abruptly after 200. The number of later writers who can be said to have made an important original contribution to scientific thought is small—although we should not forget that it was only in the fourth and third centuries B.C. that the scientists of whom that could be claimed numbered more than a handful in any one generation. Whereas most of the authors whom we have considered are only of secondary importance, exceptions must be made of Diophantus in the third century, Proclus in the fifth and Philoponus in the sixth.

Yet it is obvious that from about 200 the main effort tends more and more to be spent on preserving knowledge rather than on attempting to increase it. The commentary, as we have seen, becomes the chief vehicle of scientific writing, and although there are many exceptions, the attitude towards the texts commented on is often uncritical. The positive aspect of this admiration for earlier authors is that it insured that much scientific knowledge that would otherwise have disappeared was handed on—and this, when knowledge was constantly in danger of being lost, was no mean achievement. Yet what was thereby preserved was the *results* of Greek science: the tradition of the learned commentary could not, or at least did not, guarantee that the methods and aims of Greek science would be sustained in the continued practice of research. The spirit of inquiry was kept alive less by the scholars and commentators than by such people as the alchemists, the engineers, the herbalists and the writers of pharmacopoeia.

The history of scientific thought after 200 varies from subject to subject, but the overall picture is one of declining originality, even though this decline is arrested, and even temporarily reversed, by exceptional individuals working in particular fields. But how far can we go towards an understanding of the factors that contributed to this situation? Important social, economic and political changes are, no doubt, relevant to the problem, though how far they provide any solution to it is not clear. The relatively flourishing state of science in the second century A.D. is often associated with the comparative calm and prosperity of the Roman world under the 'good Emperors'. Yet a reading of Galen shows that even in this favourable political climate the circumstances in which a scientist found himself might be far from ideal for undisturbed research: Galen was no stranger to war, plague and famine, as well as corruption and intrigue. Conversely, although the third century was much more turbulent than the second, individuals such as the philosopher Plotinus and the mathematician Diophantus were still able to engage in intellectual studies. It cannot be stressed too often that the history of scientific thought in antiquity is the history of the ideas of a very few individuals. Attempts to establish generalizations concerning the factors that facilitated or impeded their researches are all the more hazardous since their number is so small. We can certainly pick out events that were turning points in the history of the Greco-Roman world, the division of the empire and the transfer of the capital to Constantinople in 330, the sack of Rome by Alaric in 410, the siege of 537, and the capture of Alexandria by the Arabs in 642. Among the social and economic factors at work are the shortage of manpower, the decline of agriculture, the poorness of communications and the top-heaviness of the bureaucracy. Yet when we attempt to estimate the impact of these and other circumstances on the work of the individuals we are concerned with, we must admit that we are largely reduced to guesswork.

After these general cautions, however, we may remark that some aspects of our problem provide, perhaps, a more fruitful line of inquiry. In particular, the study of religion throws light on the ideological background to the inquiry concerning nature. Here the main issue is the effect of the rise and eventual victory of Christianity, which some writers have seen as the greatest obstacle to the development of science, but which others have just as passionately denied to have any bearing whatsoever on the course of scientific thought. Two preliminary points are fundamental, first that early Christian writers disagree among themselves on many important issues, including in their attitude towards 'nature', and second that many of the beliefs of early Christianity are far from being entirely new. To take this second point first, the belief in the immortality of the soul, the relative devaluation of the body and of the material world as a whole, and the belief in the possibility of miracles, can all be paralleled in pagan religion, in Greek philosophy, or in both. Greek science coexisted with magic, superstition and irrationalism of various kinds from the very beginning. What marks Christianity out is not the particular doctrines associated with it, so much as the fact that those doctrines eventually received unprecedented state approval and support.

The attitudes adopted by Christian writers towards the inquiry concerning nature vary. For Origen, in the early third century, for instance, Greek science was not to be rejected completely. Parts of it at least can serve as a preliminary study to Christianity. As he writes in the *Epistle to Gregory* (ch 1):

> I would wish that you should draw from Greek philosophy too such things as are fit, as it were, to serve as general or preparatory studies for Christianity, and from geometry and astronomy such things as may be useful for the interpretation of the Holy Scriptures.

Augustine (354–430) is more wary. He had a great respect for Plato and the Platonic tradition, but he writes of the dangers of the curiosity for knowledge, condemning dissection in particular (*Confessions* X 35) and he makes it clear that a detailed knowledge of natural science is irrelevant:

> Nor need we be afraid lest the Christian should be rather ignorant of the force and number of the elements, the motion, order and eclipses of the heavenly bodies, the form of the heavens, the kinds and natures of animals, shrubs and stones . . . It is enough for the Christian to believe that the cause of all created things, whether heavenly or earthly, whether visible or invisible, is none other than the goodness of the Creator, who is the one true God. (*Enchiridion* III 9.)

But a much more openly hostile attitude towards the inquiry concerning nature is expressed by other writers, particularly the Latin fathers Tertullian (about 200 A.D.) and Lactantius (early fourth century) though from the Greek tradition we may add Cosmas Indicopleustes of Alexandria, for instance, from the sixth century. To take just one example, Tertullian, who believes that the Greek philosophers plagiarized their ideas from the Old Testament prophets and then distorted what they had learnt, undertakes to refute them in several of his works. He criticizes them for the uncertainty of their opinions and for their constant disagreements with one another. He sees philosophy as a threat to the Christian faith and rejects it categorically:

> What then has Athens to do with Jerusalem, the Academy with the Church, the heretic with the Christian? Our instruction comes from the Porch of Solomon who himself taught us that the Lord is to be sought in the simplicity of one's heart . . . We have no need of curiosity after Jesus Christ, nor of research after the gospel. When we believe, we desire to believe nothing more. For we believe this first, that there is nothing else that we should believe. (*On Prescriptions against Heretics* ch 7.)

One point on which the early Christian writers were, in general, agreed is that the profoundest truths come neither from observation, nor from reason, but from divine revelation.

For the faithful, empirical inquiry is unnecessary, a distraction from the practice of his religion and possibly a source of dangerous heresy. Unlike some other Christian beliefs and attitudes, this doctrine, if rigidly adhered to, meant the end of scientific research. Admittedly the idea that truth depends on divine revelation is not confined to Christianity. But the very success of that religion insured that this doctrine acquired a prominence and an influence that it had never had before.

Moreover after the adoption of Christianity as the official religion of the Emperor, pagan philosophers and scientists had not only an unfavourable climate of opinion to contend with, but also, on occasions, legal sanctions. The most famous instance is Justinian's edict of 529 closing the Academy at Athens and forbidding pagans to teach: if they were not baptized, they were liable to exile and confiscation of property. But although Justinian is often thought to have thereby administered the *coup de grâce* to pagan learning, this is to exaggerate the effectiveness of his measure. It certainly had an immediate impact on, among others, Simplicius, who, together with other leading neo-Platonists, left Athens to go to the court of King Chosroes of Persia, in whom they hoped to find a Platonic philosopher-king. Yet after perhaps three years in the East Simplicius seems to have returned to Athens. Although confiscations of property had begun, there is some evidence[1] that the Academy survived in some form, and we certainly hear of other pagan philosophers who continued to teach in other cities, for example Olympiodorus in the latter part of the sixth century at Alexandria.

Legal measures were never as successful in controlling men's minds as their initiators hoped. The way in which Christianity survived the persecutions of the pagan Emperors, including the apostate Emperor Julian in the fourth century, is eloquent testimony to this. It is doubtful whether, once the Church was in power, it could have stamped out scientific inquiry completely even if it had consistently set out to do so. And we have seen that in the sixth century one of those who helped to keep science alive, John Philoponus, was himself (whether

[1] For a recent reevaluation of the evidence, see A. Cameron, 'The Last Days of the Academy at Athens', *Proceedings of the Cambridge Philological Society*, vol. 195, 1969, pp 7 ff.

by birth or conversion we do not know) a Christian. Yet if neither the polemic of such writers as Tertullian nor the laws of Justinian totally eradicated the inquiry concerning nature, both were nevertheless symptomatic of an age that became increasingly hostile to that inquiry. The men who engaged in what we should call science had always been a tiny minority who faced the indifference of the mass of their contemporaries at every period. But in late antiquity the triumph of Christianity both symbolized, and itself contributed to, a deterioration in an already unfavourable climate of opinion.

With the Christian Church, religion became established institutionally in ancient society in a way in which science never did. Scientific research received, as we have seen, only sporadic and uncertain support from the rulers of states: this remains true as a generalization, even if there are some exceptions, notably the Alexandrian Museum, which continued in existence right down to the fifth century A.D. To be sure, schools of medicine such as that of Alexandria (this too survived into the Byzantine period) and institutions of higher learning such as the Academy and the Lyceum, provided from an early date a focal point for those who were interested in what we should describe as scientific problems. Yet important as those schools were, their primary *raison d'être* was an educational one. Their survival depended on this function, which they continued to fulfil whether or not any of their members carried on original scientific research. The contrast with religion is instructive. From about 250 onwards the Church grew very rapidly in power and wealth, until indeed it came to be a major factor contributing to the economic difficulties of the later Roman Empire. As Jones has put it:

> Economically the church was an additional burden, which steadily increased in weight, on the limited resources of the empire. The huge army of clergy and monks were for the most part idle mouths, living upon offerings, endowments and state subsidies.[1]

Moreover the Church soon came to represent both the most prestigious and the most lucrative career. We have remarked

[1] A. H. M. Jones, *The Later Roman Empire 284–602*, Blackwell, Oxford, 1964, vol 2, p 933.

that while most of those who engaged in scientific research enjoyed private means, many earned a living as doctors or teachers. Yet in the sixth century, to quote Jones again, the bishop of Anastasiopolis enjoyed a salary that was six times that of the public doctor at Antinoopolis, and five times that of professors of rhetoric or grammar at Carthage. The ancient writers themselves, both Christian and pagan, often disclaim any motive of financial reward: money, they suggest, is less important than virtue and honour. But however we evaluate these statements, the fact is that from the third century onwards both criteria were most fully satisfied by a career in the Church.

The story of scientific activity after 200 is, as we have seen, one not of sudden cessation, but of gradual decline. Scepticism concerning the possibility of discovering the true causes of phenomena, and deference to the authority of earlier writers, were certainly not new features in Greek thought. They are present from the very beginnings of Greek science, in for example the statement of the sixth century B.C. poet-philosopher Xenophanes of Colophon that 'no man knows or ever will know the clear truth about the gods and about everything I speak of' (fragment 34), and in the respect that the early Pythagoreans had for the founder of their school. But in late antiquity a note of desperation is sounded. In astronomy, for example, Proclus and Philoponus record the ideas of Ptolemy but register their doubts not only about his hypotheses, but about the possibility of accounting for the movements of the heavenly bodies at all. And these two men were, each in his own day, the foremost representatives of science. From the beginning, the chief driving force of Greek science had been curiosity, the desire for knowledge for its own sake. But in late antiquity the spirit of inquiry had to contend first with the belief in the superior wisdom of the great writers of the past, second with the growing feeling of the impossibility of achieving true understanding concerning the causes of natural phenomena, and third with the direct attack of those, such as the Christian writers we have mentioned, who preferred revelation to reason and sensation and put faith above knowledge.

In time, even the desire to preserve the knowledge won in the past atrophied. Institutions of higher education continued right down into the Byzantine period. Yet the learning transmitted from one generation to the next is diminished and distorted. Already in the second century Galen had complained that his own contemporaries were ignorant of much that the ancients had known:

> I find that many things that have been perfectly demonstrated by the ancients are not understood by many people nowadays because of their ignorance—indeed because through laziness they do not even try to comprehend them. (*On the Natural Faculties* III ch 10.)

Thereafter the commentary and abridgement play decisive roles both in transmitting and at the same time narrowing scientific knowledge. One example has already been mentioned. Galen's own works formed the main source for the vast *Medical Collection* of Oribasius: but Paul of Aegina, three centuries later, considered this work too long and composed a shorter compendium of medical knowledge of his own. The selection or abridgement, which may originally be intended as an elementary guide to the subject, ends by supplanting the original work itself.

The level of knowledge declines much more rapidly in the Roman West than in the Greek East. In the West what survived was mostly contained in handbooks that owed their popularity mainly to the curiosities they recorded. Pliny's *Natural History* (c. A.D. 75) and Aulus Gellius' *Attic Nights* (second century) are followed by Solinus' *Collection of Remarkable Facts* (third or fourth century). Some knowledge of the elementary parts of Aristotle's logic and of elementary mathematics is transmitted through Boethius (late fifth century). Some idea of Platonic cosmology was preserved through the Latin commentary on the *Timaeus* by Chalcidius (about A.D. 400) and a hotch-potch of information was contained in the popular works of such writers as Macrobius (for example the *Commentary on the Dream of Scipio*) and Martianus Capella (*The Marriage of Mercury and Philology*). Although Boethius and others used some Greek sources directly, the compilers of handbooks tended to rely more and more on other handbooks, and we find the same fantastic stories repeated by one author after another. The low level of knowledge in the West in the seventh century is illustrated by the fact that Isidore of Seville, who had the reputation of being one of the most learned individuals of the age, believed that the stars are lit by the sun (*On the Nature of Things*, ch 24).

In the East, by contrast, much more of Greek science was preserved. The tradition of scholarship kept alive a knowledge of the ancient scientific texts even when the scholarly commentators themselves did not attempt to engage in original scientific research on their own account. Much of what we know about Greek science is due to this tradition: we owe the preservation of much of Apollonius, for example, to his sixth-century editor, Eutocius. After the fall of Alexandria to the Arabs (642), knowledge of medicine, biology, astronomy and mathematics spread through the Arab world, and from about the middle of the ninth century the Arabs produced scholars such as the astronomer and geometer Thabit ben Qurra and the polymath Al Kindi who themselves entered into the debate on various problems in natural science.

The history of the contributions of the Arabs and the transmission of scientific knowledge back into the West cannot be discussed here. But it is relevant to note first that knowledge of Greek science never completely died out, and secondly that the rebirth of scientific inquiry involved not only a rediscovery of Greek learning but also a return to the original aims and methods of the greatest Greek scientists. The authority of Aristotle, Ptolemy and Galen, so long unchallenged in matters of natural philosophy, had in due course to be overthrown, and, more fundamentally, the notion of *authority* itself had to be reappraised. But to reject authority in favour of first hand observation and research was, after all, true to the spirit of the earliest Greek scientists themselves. The continuities between ancient science and the beginnings of modern science in the sixteenth and seventeenth centuries are as important as their discontinuities. I have already mentioned what Copernicus' *De Revolutionibus* owes, in form and method as well as content, to Ptolemy (pp 129 f). The ambivalence of the 'moderns' towards the 'ancients' is seen even more clearly in a second work that appeared in the same year (1543) as the *De Revolutionibus*, namely Vesalius' *De Humani Corporis Fabrica.* Vesalius' admiration for Galen is equalled only by the zeal with which he exposes and castigates his errors—although, as is well known, there are many occasions when he repeats Galen's mistakes. But the important point is that Vesalius' criticisms are themselves the result of his rediscovery, and extension, of the method that Galen himself had used, namely dissection. A third, more complex, example among the many others that could be given is the revival of interest in Plato and the Pythagoreans, to whom both Galileo and Kepler looked as ancient models and authorities in their own search for mathematical order in physics and cosmology.

* * *

It is no more possible to give a summary assessment of more than 1000 years of Greek science than it would be to try to do so for Greek literature. Nevertheless we may take stock of some of the findings of our study in some concluding remarks. In one sense Greek science may be considered a failure. The conditions needed to insure the continuous growth of science did not exist, and were never created, in the ancient world. There were certainly doctors, architects and engineers who recognized the practical importance of some aspects of their theoretical inquiries. Yet their efforts were uncoordinated, and no systematic attempt to explore the practical applications of science was made. Of the possible *raisons d'être* of science, the idea that it could be of practical use, while not totally absent, took second place to the idea that the study of nature contributed to knowledge and understanding—which are valuable for their own sakes. A large part of ancient natural science never fully emancipated itself from philosophy: but to put it that way is to speak from a modern, not an ancient, standpoint. To the ancients, philosophy generally included physics or the inquiry concerning nature as one of its three main branches, and the chief motive for that inquiry was the philosophical one in the literal sense of the 'love of wisdom'. So we find some of the doctors and architects assimilating their studies to philosophy, and if this was in part because of the superior social position and prestige of philosophy, it also reflects the belief that their principal goal was knowledge. Science was less a means to an end, than an end in itself. The life devoted to study or 'contemplation' is the supremely happy life. Knowledge is its own reward and correspondingly less attention was paid to the idea of the benefits that might accrue from its application to practical purposes.

The tendency to value material below moral and intellectual goods is common in the writings of ancient moral philosophers, and although one may doubt how far their opinions were typical of ancient society as a whole, the lack of any sustained attempt to justify scientific inquiry in terms of the increased material prosperity to which it might lead is one of the most striking differences between the ancient view and that of the nineteenth and twentieth centuries. But it is not merely the case that ancient science reflects ancient values. Science itself was, in at least three ways, not morally neutral. First Plato's idea (*Timaeus* 47 bc), that contemplation of the orderly movements of the heavenly bodies helps us to regularize the disorderly movements of our own soul, is repeated in a rather simpler form by later writers. Ptolemy, for instance, believed that astronomy improves men's characters. Secondly, there were writers who, without necessarily claiming that science makes men better, maintained that nature is orderly, beautiful and good (whether or not they also postulated an intelligent and benevolent deity responsible for its design). Thirdly, even among those who rejected in the strongest possible terms the idea that the world is the product of design, there were some who justified the study of nature on what are, broadly speaking, ethical grounds. Thus the Epicureans shared the view of their opponents, the Stoics, that some knowledge of natural phenomena is necessary to insure the peace of mind without which a man cannot be truly happy.

But if the study of nature is often linked to, or even part of, moral philosophy, this generalization, like so many others, needs qualification. We happen to have good evidence concerning the views of the Epicureans, the early Stoics, and such men as Ptolemy and Galen, on the place that the inquiry concerning nature has in a scheme of values. In many other cases, however, we have no definite or reliable information on this point. Physics, mathematics and biology were often conceived as parts of a comprehensive philosophy. Yet they could be, and often were, studied as independent disciplines with no commitment on the part of the researcher to any overall cosmological or ethical theses. Much of the best work in mathematics, in particular, was done by men such as Archimedes and Apollonius who we have no reason to believe treated their scientific investigations as part of a systematic philosophy in the way in which the Stoics and Epicureans did.

Moreover there were those who deliberately dissociated themselves from the philosophers, though not on ethical grounds, so much as on epistemological or methodological ones. Already in the period before Aristotle there had been doctors who rejected the philosophers' methods of inquiry as mere speculation[1] and later medical theorists, particularly among the Empiricists, sometimes did the same. Criticism of the philosophers for placing too much reliance on abstract argument is part of the continuing debate on the roles of reason and sensation, argument and experience, which runs through the work of many ancient scientists. Thus Philo remarks, in connection with the study of ballistics, that 'it is not possible to arrive at a complete solution of the problems involved merely by reason and by the methods of mechanics . . . many discoveries can be made only as a result of trial' (p 99), and again Philoponus, discussing the movement of falling bodies, claims that his own view 'can be established by what is actually observed more powerfully than by any sort of demonstration by arguments' (p 160).

The social and intellectual framework within which ancient scientists worked differed in certain fundamental respects from that of their modern counterparts. There was no acknowledged place in ancient thought, or in ancient society, for science, or for the scientist, as such. The investigators performed different social roles as doctors or architects or teachers. Disagreements concerning the 'inquiry about nature' were not merely a matter of the differing motives and philosophical allegiances of the individuals concerned, but directly concerned the types of question to be investigated. For some nature was everywhere purposeful, but others questioned teleology and final causes or rejected them outright, while still seeking order and regularities expressible as general laws.

But despite many important differences, the relevance of the

[1] See *Early Greek Science*, pp 59 ff.

work of the ancients to what we mean by science remains. This is true in one obvious way in that in such fields as optics, statics, astronomy and anatomy, the ancients achieved certain positive results (admittedly in the elementary parts of those disciplines) which provided a basis on which later scientists could build directly. But even more important was the creation, elaboration and exemplification of models of the inquiry concerning nature itself. Two key methodological principles, the application of mathematics to the investigation of natural phenomena, and the notion of deliberate empirical research, go back to the earlier period that culminates in Aristotle. What the later period we have considered in this study provides is, above all, examples of the application of these principles in practice. Euclid's *Elements* was the chief model of an axiomatic, deductive system. Archimedes in statics and hydrostatics, and Apollonius and Ptolemy in astronomy, represent the most successful attempts to bring mathematical methods to bear on the discussion of physical problems. As for the idea of research, we find a powerful statement of this in Erasistratus, and among others such as Theophrastus, Strato, Herophilus and Hipparchus, who might be mentioned, Ptolemy and Galen provide excellent examples of the practice of *historiā*. Both men not only carried out extensive programmes of observation, but also conducted deliberate tests, Ptolemy in optics, and Galen in his investigations of the nervous system and other vital functions.

The criticism is often made that the fatal shortcoming of Greek science was the failure to appreciate the importance of experimentation. But that is an oversimplification. It is true that the use of the experimental method is confined to certain problems and to certain individuals, but the same may also be said of the idea of the mathematization of physics. Here too the principle was known, and it is not hard to identify, with the benefit of hindsight, the opportunities for its application that were missed. In neither case is there a fundamental difference in kind, however great the differences in degree, between the methods of ancient and modern science. But both these shortcomings in ancient science reflect, and were aggravated by, the more basic organizational weakness to which I have alluded, the fact that the conditions needed to insure the continuous growth of science never existed in the ancient world. The relative isolation of those who engaged in scientific investigation acted as an obstacle to the systematic application of methodological ideas and was a constant threat to the continuity of inquiry in most fields of science. Thus we know of no significant contribution to the study of dynamics between Hipparchus, in the second century B.C., and Philoponus and Simplicius in the sixth century A.D.—although both these later writers generally provide full information concerning the work of their predecessors. Optics, botany and embryology too suffered from long periods of stagnation (in several cases the work done in the fourth or third century B.C. was never surpassed) and the same can be said of most other branches of investigation with the exception of elementary mathematics, astronomy and medicine. Moreover, even within the same generation the lack of communication between men who were interested in different aspects of the inquiry concerning nature can be illustrated by comparing the statements that Epicurus made on problems in astronomy with the theories of some of his own contemporaries. Given the conditions under which scientists worked, it is hardly surprising that some of the most important theories, discoveries and methods of ancient science were sometimes ignored or not followed up with any vigour. Yet the neglect that some of the most important ideas produced by the ancients suffered from in antiquity does not diminish the value of those ideas in themselves. The weakness of the social and ideological basis of ancient science becomes more obvious in the decline we have outlined in this chapter. But when scientific investigation was revived in the West, it was a genuine rebirth, not merely in that the work of the great ancient scientists was rediscovered, but also and more particularly in that there was a return to the spirit of inquiry of ancient science and to the models of method that it provided.

Select Bibliography

I Sources: Texts and Translations

A *General*

A Source Book in Greek Science, edited by M. R. Cohen and I. E. Drabkin (second edition, Cambridge, Mass., Harvard University Press, 1958). This does not include cosmology, but otherwise provides a full selection of the most important passages in translation with a useful bibliography.

B *The Lyceum after Aristotle*

(i) Theophrastus. Editions with translations: W. D. Ross and F. H. Fobes, *Theophrastus Metaphysics* (Oxford, Clarendon Press, 1929); E. R. Caley and J. F. C. Richards, *Theophrastus On Stones* (Columbus, Ohio, Ohio State University, 1956); Sir Arthur Hort, *Theophrastus Enquiry into Plants*, 2 vols, Loeb edition (Cambridge, Mass., Harvard University Press; London, Heinemann, 1916); R. E. Dengler, *Theophrastus De Causis Plantarum, Book One* (Philadelphia, 1927). Otherwise the latest complete text of the scientific works is that of the Teubner edition (Leipzig, ed. F. Wimmer, 1854–62).

(ii) Strato: *Straton von Lampsakos*, edited by F. Wehrli as vol 5 of *Die Schule des Aristoteles* (2nd ed, Basel, Schwabe, 1969). Some texts are translated in H. B. Gottschalk, *Strato of Lampsacus: some texts* (in *Proceedings of the Leeds Philosophical and Literary Society*, Literary and Historical Section, vol XI (1964–6), Part VI, 1965).

C *Epicureans and Stoics*

(i) Epicurus and Lucretius: *Epicuro Opere*, ed. G. Arrighetti (Torino, Einaudi, 1960); *Epicurus* (with translation), C. Bailey (Oxford, Clarendon Press, 1926); *Lucretius* (with translation), C. Bailey, 3 vols (Oxford, Clarendon Press, 1947).

(ii) Stoics: *Stoicorum Veterum Fragmenta*, 4 vols, ed. H. von Arnim (Leipzig, Teubner, 1905–24). Many of the passages relevant to Stoic physics are translated in S. Sambursky, *The Physics of the Stoics* (London, Routledge and Kegan Paul, 1959).

D *Hellenistic Mathematics and Astronomy*

With the exception of Aristarchus, edited by Heath (see below), the latest comprehensive editions of the major Hellenistic mathematicians and astronomers are those in the Teubner (Leipzig) series: Euclid (ed. I. L. Heiberg and H. Menge, 1883–1916), Archimedes (second edition, I. L. Heiberg, 1910–15), Apollonius (ed. I. L. Heiberg, 1891–3) and Hipparchus (*In Arati et Eudoxi Phaenomena*, ed. C. Manitius, 1894). The best translations are still in most cases those of T. L. Heath: *The Thirteen Books of Euclid's Elements*, 3 vols (Cambridge, University Press, 1908); *Aristarchus of Samos* (with text, Oxford, Clarendon Press, 1913); *The Works of Archimedes* (Cambridge, University Press, 1912; Dover Books (no date)); *Apollonius of Perga* (Cambridge, University Press, 1896, W. Heffer, 1961). See also E. J. Dijksterhuis, *Archimedes* (Copenhagen, Munksgaard, 1956) and D. R. Dicks, *The Geographical Fragments of Hipparchus* (University of London, Athlone Press, 1960; New York, Oxford University Press).

There are also selections of texts in translation in the 2 volume Loeb edition, *Greek Mathematical Works*, ed. I. Thomas (Cambridge, Mass., Harvard University Press; London, Heinemann, 1939–41) and T. L. Heath, *Greek Astronomy* (London, Dent, 1932; New York, AMS Press, Inc.).

E *Hellenistic Biology and Medicine*

No adequate edition of the fragments of Herophilus and Erasistratus is available. Some fragments of Herophilus were collected and commented on by K. F. H. Marx, *De Herophili celeberrimi medici vita* (Göttingen, 1840). There are selections of passages in translation in J. F. Dobson, 'Herophilus of Alexandria' and 'Erasistratus', *Proceedings of the Royal Society of Medicine*, no. 18, 1924–5, pp 19–32 and no. 20, 1926–7, pp 825–32. Our two most important ancient sources are Celsus (see W. G. Spencer, *Celsus De Medicina*, 3-vol Loeb edition, Cambridge, Mass., Harvard University Press; London, Heinemann, 1935–8) and Galen (see below, I H).

F *Applied Mechanics and Technology*

There is a Loeb edition of Vitruvius (F. Granger, *Vitruvius On Architecture*, 2 vols, Cambridge, Mass., Harvard University Press; London, Heinemann, 1931–4) and Teubner editions of Vitruvius (F. Krohn, 1912) and Hero (W. Schmidt and others, 1899–1914: vol 1 also contains Philo's *Pneumatics*).

Translations of many important passages are to be found in A. G. Drachmann, *Ktesibios, Philon and Heron, A Study in Ancient Pneumatics* (Copenhagen, Munksgaard, 1948), the same author's *The Mechanical Technology of Greek and Roman Antiquity* (Copenhagen, Munksgaard, 1963) or in E. W. Marsden, *Greek and Roman Artillery, Technical Treatises* (Oxford, Clarendon Press, 1971).

G *Ptolemy*

Texts: *Claudii Ptolemaei Opera quae exstant omnia*, 3 vols but still incomplete, ed. I. L. Heiberg and others (Leipzig, Teubner, 1898–

1952); *L'Optique de Claude Ptolémée*, ed. A. Lejeune (Louvain, Publications universitaires de Louvain, 1956).

Translations: R. Catesby Taliaferro, *Ptolemy, The Almagest* (Chicago, Encyclopedia Britannica, 1952); F. E. Robbins, *Ptolemy, Tetrabiblos*, Loeb edition (Cambridge, Mass., Harvard University Press; London, Heinemann, 1940).

H *Galen*

Texts: The most recent comprehensive edition is C. G. Kühn, *Claudii Galeni Opera Omnia* 20 vols in 22 (Leipzig, Cnobloch, 1821–33). This is gradually being superseded by the Corpus Medicorum Graecorum edition (various editors, Leipzig, Teubner, in progress since 1914). Some treatises have also been edited in *Claudii Galeni Pergameni Scripta Minora*, 3 vols, ed. J. Marquardt and others (Leipzig, Teubner, 1884–1893).

Translations: The most important English translations are: C. Singer, *Galen, On Anatomical Procedures* (Oxford, University Press, 1956); *Galen, On Anatomical Procedures, The Later Books*, trans. W. L. H. Duckworth, ed. M. C. Lyons and B. Towers (Cambridge, University Press, 1962); A. J. Brock, *Galen, On the Natural Faculties*, Loeb ed. (Cambridge, Mass., Harvard University Press; London, Heinemann, 1916). M. T. May, *Galen, On the Usefulness of the Parts of the Body*, 2 vols (Cornell University Press, Ithaca, New York, 1968).

I *Science after* 200 *A.D.*

There are Teubner editions of several of the more important texts: Iamblichus, *De Communi Mathematica Scientia* (ed. N. Festa, 1891), Diophantus, *Arithmetica* (ed. P. Tannery, 1893), Proclus, *Hypotyposis Astronomicarum Positionum* (ed. C. Manitius, 1909) and two of Philoponus' works, *De Opificio Mundi* (ed. W. Reichardt, 1897) and *De Aeternitate Mundi contra Proclum* (ed. H. Rabe, 1899). Pappus, *Collectio Mathematica*, has been edited by F. Hultsch (Berlin, Weidmann, 3 vols, 1876–8). The commentaries on treatises of Aristotle by Philoponus and Simplicius have been edited in the Berlin Academy series *Commentaria in Aristotelem Graeca* (Berlin, Reimer, 1882–1909).

Translations of many of the most important passages will be found in S. Sambursky, *The Physical World of Late Antiquity* (London, Routledge and Kegan Paul, 1962). See also, T. L. Heath, *Diophantus of Alexandria* (second edition, Cambridge, University Press, 1910, Dover books, 1964).

II Secondary Reading

A *General*

The most important works in English are:

S. Sambursky, *The Physical World of the Greeks* (trans. M. Dagut, London, Routledge and Kegan Paul, 1956; New York, Humanities Press, Collier-Macmillan (paper) 1956)

O. Neugebauer, *The Exact Sciences in Antiquity* (second edition, Providence, R.I., Brown University Press, 1957; Harper torchbooks, 1962)

M. Clagett, *Greek Science in Antiquity* (London, Abelard-Schuman, 1957; New York, Collier-Macmillan (paper))

B. Farrington, *Greek Science* (revised one vol edition, London, Penguin Books, 1961; Baltimore, Md., Penguin Books)

See also:

G. Sarton, *A History of Science*, 2 vols (London, Oxford University Press; Cambridge Mass., Harvard University Press, 1953–59)

B *The Lyceum after Aristotle*

No comprehensive study of Theophrastus' scientific achievements is available in English (the standard work is P. Steinmetz, *Die Physik des Theophrastos von Eresos*, Bad Homburg, Verlag M. Gehlen, 1964), but the editions cited in I B above are helpful.

For Strato, consult the work of H. B. Gottschalk, cited above I B.

C *Epicureans and Stoics*

C. Bailey, *The Greek Atomists and Epicurus* (Oxford, Clarendon Press, 1928; New York, Russell and Russell)

S. Sambursky, *The Physics of the Stoics* (London, Routledge and Kegan Paul, 1959)

D. J. Furley, *Two Studies in the Greek Atomists* (Princeton, N.J., Princeton University Press, 1967)

D *Hellenistic Mathematics and Astronomy*

The best available general study of Greek mathematics is still T. L. Heath, *A History of Greek Mathematics*, 2 vols (Oxford, Clarendon Press, 1921)

Apart from T. L. Heath, *Aristarchus of Samos*, and O. Neugebauer, *The Exact Sciences in Antiquity* (already cited, I D and II A, above) the following studies are fundamental for Hellenistic astronomy:

O. Neugebauer, 'The History of Ancient Astronomy; Problems and Methods', *Journal of Near Eastern Studies*, no. 4 (1945), pp 1–38;

'The equivalence of eccentric and epicyclic motion according to Apollonius', *Scripta Mathematica*, no. 24 (1959), pp 5–21

B. L. van der Waerden, *Science Awakening* (trans. A. Dresden, Groningen, Noordhoff, 1954; New York, Oxford University Press, 1961; John Wiley and sons (paper))

On Greek astronomical instruments, consult:

D. R. Dicks, 'Ancient astronomical instruments', *Journal of the British Astronomical Association*, no. 64 (1954), pp 77–85

D. J. de S. Price, 'Precision instruments: to 1500' in *A History of Technology*, vol 3, ed. C. Singer and others (Oxford, Clarendon Press, 1957), pp 582–619

E *Hellenistic Biology and Medicine*

L. Edelstein, *Ancient Medicine* (Baltimore, Johns Hopkins Press, 1967) contains articles on, for example, the development of anatomy, and Greek medical sects. Two other important articles are:

O. Temkin, 'Greek medicine as science and craft', *Isis*, no. 44 (1953), pp 213–25

L. G. Wilson, 'Erasistratus, Galen and the Pneuma', *Bulletin of the History of Medicine*, no. 33 (1959), pp 293–314

F *Applied Mechanics and Technology* (see also I F above)

A. G. Drachmann, *Ancient Oil Mills and Presses* (Copenhagen, Levin and Munksgaard, 1932)

C. Singer, E. J. Holmyard, A. R. Hall, T. I. Williams (ed.), *A History of Technology*, vols 1–3 (Oxford, Clarendon Press, 1954–57)

R. J. Forbes, *Studies in Ancient Technolcgy* 9 vols have so far appeared, some in a second edition (Leiden, Brill, in progress since 1955)

L. A. Moritz, *Grain-Mills and Flour in Classical Antiquity* (Oxford, Clarendon Press, 1958)

M. I. Finley, 'Technical innovation and economic progress in the ancient world', *Economic History Review*, 2nd ser., no. 18 (1965), pp 29–45

E. W. Marsden, *Greek and Roman Artillery, Historical Development* (Oxford and New York, Clarendon Press, 1969)

G *Ptolemy*

The best brief account of Ptolemy's astronomy is that in Appendix 1 of O. Neugebauer, *The Exact Sciences in Antiquity* (see II A above)

See also:

L. O. Kattsoff, 'Ptolemy and scientific method', *Isis*, no. 38 (1947), pp 18–22

T. S. Kuhn, *The Copernican Revolution* (Cambridge, Mass., Harvard University Press, 1957)

H *Galen*

No good general account of Galen is available. See however:

G. Sarton, *Galen of Pergamon* (Lawrence, Kansas, University of Kansas Press, 1954)

D. Fleming, 'Galen on the motions of the blood in the heart and lungs', *Isis*, no. 46 (1955), pp 14–21

R. E. Siegel, *Galen's system of physiology and medicine* (Basel and New York, S. Karger, 1968); *Galen on sense perception* (Basel and New York, S. Karger, 1970)

I *Science after* 200 *A.D.*

A. C. Crombie, *Augustine to Galileo*, 2 vols (second edition, London, Heinemann, 1959; Peregrine Books (paper) 1969)

S. Sambursky, *The Physical World of Late Antiquity* (London, Routledge and Kegan Paul, 1962); 'Conceptual developments and modes of explanation in later Greek scientific thought' in *Scientific Change*, ed. A. C. Crombie (London, Heinemann, 1963), pp 61–78; 'Plato, Proclus, and the limitations of science', *Journal of the History of Philosophy*, no. 3 (1965), pp 1–11

W. H. Stahl, *Roman Science* (Madison, University of Wisconsin Press, 1962)

J *Methodology*

L. Edelstein, 'Recent trends in the interpretation of ancient science', *Journal of the History of Ideas*, no. 13 (1952), pp 573–604, reprinted in *Roots of Scientific Thought* (ed. P. P. Weiner and A. Noland, New York, Basic Books, 1957), pp 90–121

A. Wasserstein, 'Greek scientific thought', *Proceedings of the Cambridge Philological Society*, no. 188 (n.s. 8) (1962), pp 51–63

G. E. R. Lloyd, 'Experiment in early Greek philosophy and medicine', *Proceedings of the Cambridge Philological Society*, no. 190 (n.s. 10) (1964), pp 50–72

Index

INDEX

INDEX

INDEX

INDEX

09 01 409779 01 10
LLOYD. GEOFFREY ERNEST RICHARD
GREEK SCIENCE AFTER ARISTOTLE
(1) 1973 9-509.495L793G
175275
YA
509.495 L793 g
175275
LLOYD
GREEK SCIENCE AFTER ARISTOTLE
JAN 27 '86
DEC 11 1989
DISCARD
FEB 26 1990
QUEENS BOROUGH PUBLIC LIBRARY
CENTRAL LIBRARY
89-11 Merrick Boulevard
Jamaica, New York 11432
Ya
DISCARD
All items may be borrowed for 28 days and are due on latest date stamped on card in pocket.
A charge is made for each day, including Sundays and holidays, that this item is overdue.
SORRY NO RENEWALS